Lecture Notes in Mathematics

Edited by A. Dold and B. Eckmann

1075

Hideyuki Majima

Asymptotic Analysis for Integrable Connections with Irregular Singular Points

Springer-Verlag
Berlin Heidelberg New York Tokyo 1984

Author

Hideyuki Majima
Department of Mathematics, Faculty of Science, University of Tokyo
Hongo, Tokyo 113, Japan

AMS Subject Classification (1980): 58A17, 35C20; 32L10

ISBN 3-540-13375-5 Springer-Verlag Berlin Heidelberg New York Tokyo
ISBN 0-387-13375-5 Springer-Verlag New York Heidelberg Berlin Tokyo

Printed in Germany

Printing and binding: Beltz Offsetdruck, Hemsbach / Bergstr.
2146/3140-543210

Dedicated to the Author's Grand-Professor Masuo Hukuhara
on his 77th Birthday

Abstract

Using strongly asymptotic expansions of functions of several variables, we prove existence theorems of asymptotic solutions to integrable systems of partial differential equations of the first order with irregular singular points under certain general conditions. We also prove analytic splitting lemmas for completely integrable linear Pfaffian systems. Moreover, for integrable connections with irregular singular points, we formulate and solve the Riemann-Hilbert-Birkhoff problem, and prove analogues of Poincaré's lemma and de Rham cohomology theorem under certain general conditions.

Key words and phrases. Strongly asymptotically developable, strongly asymptotic expansions of functions of several variables, sheaf of germs of strongly asymptotically developable functions, real blow-up of a complex manifold along a normal crossing divisor, vanishing theorem, integrable system of partial differential equations of the first order with regular and irregular singular points, existence theorems of asymptotic solutions to integrable systems with singular points, splitting lemmas for completely integrable Pfaffian systems with singular points, Stokes multipliers, Stokes phenomena, Riemann-Hilbert-Birkhoff problem, Integrable connections with regular and irregular singular points, ∇-de Rham complex for integrable connection ∇, ∇-Poincaré lemma, ∇-de Rham cohomology theorem.

Preface

The purpose of this paper is to lay a foundation of studies on integrable connections, i.e., locally speaking, completely integrable Pfaffian systems or systems of partial differential equations of the first order, with irregular singular points. Using strongly asymptotic expansions of functions of several variables, we prove existence theorems of asymptotic solutions to integrable systems of partial differential equations under certain general conditions. Moreover, we formulate and solve the Riemann-Hilbert-Birkhoff problem, and provide analogues of Poincaré's lemma and de Rham cohomology theorem for integrable connections with irregular singular points under certain general conditions.

This study is done within the framework of our theory of strongly asymptotic developability of functions of several variables, which will be reviewed in Part I. Further we shall provide, in Part I, the notation used in the following Parts II-IV. We shall prove in Part II existence theorems of asymptotic solutions to integrable systems of partial differential equations of the first order with irregular singular points under certain general conditions. We shall also prove splitting lemmas for completely integrable Pfaffian systems with irregular singular points. These theorems are main results of this paper by themselves, but also preliminaries to Parts III and IV. In Part III, we shall describe the so-called Stokes phenomena and formulate the Riemann-Hilbert-Birkhoff problem. We shall solve this problem using a vanishing theorem of noncommutative case stated in Part I. In the last Part, we shall provide analogues of Poincaré's lemma and de Rham cohomology theorem for integrable connections with irregular singular points under certain general conditions. A vanishing theorem of commutative case will be utilized here.

We shall provide a general introduction in Part 0 in order to give an overview of the whole. Moreover, a historical introduction will be given in each Part.

The reader can start from any one of four Parts according to his interest.

Originally, the author prepared two preprints [51, 54] for this study (cf. Proceedings, [50, 52, 53]). However, he reorganized them because the first two Parts have inseparable relations with each of the last two Parts (see Chart).

The author would like to express his gratitude to his three professors: to Professor R. Gérard and his colleagues at Strasbourg and Metz in France for their hospitality; to Professor Y. Sibuya at Minnesota in the U.S.A. for many fruitful discussions, valuable suggestions and his kind offices; and to Professor T. Kimura at Tokyo in Japan for perpetual encouragement and support.

Hideyuki Majima *

* Partially supported by the "Sakkokai Foundation" and partially supported by a grant from the National Science Foundation of the U.S.A.

Table of Contents

Chart

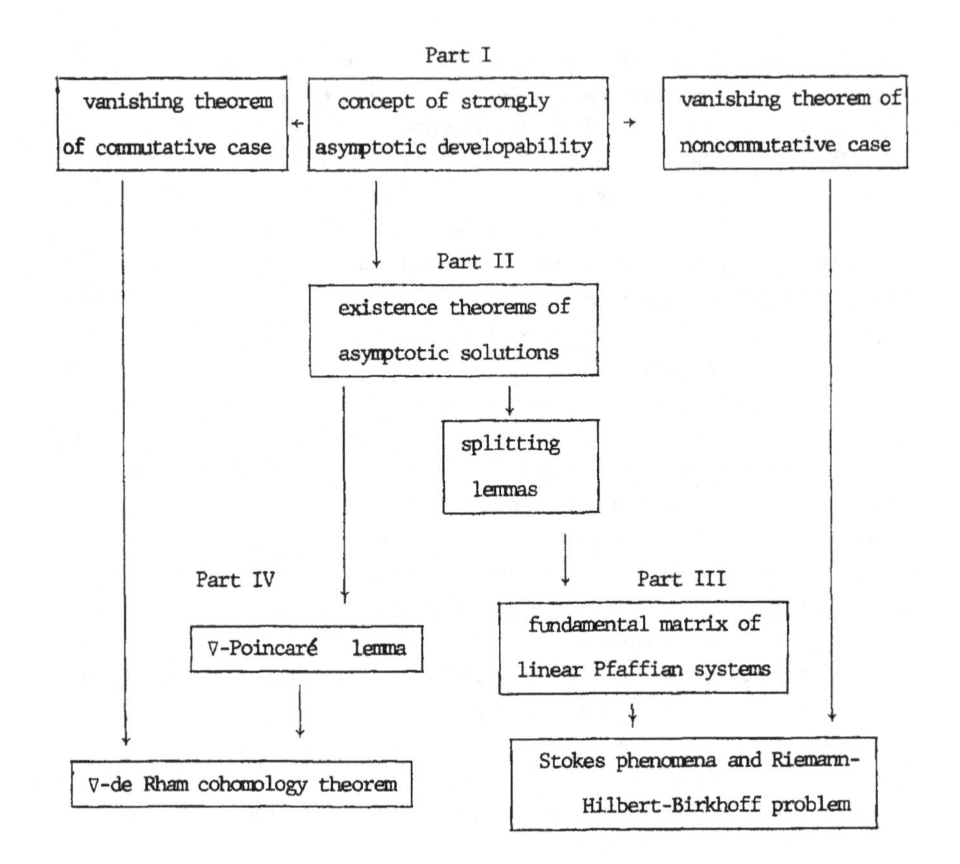
Part I
vanishing theorem
of commutative case
concept of strongly
asymptotic developability
vanishing theorem of
noncommutative case
Part II
existence theorems of
asymptotic solutions
splitting
lemmas
Part IV
∇-Poincaré lemma
Part III
fundamental matrix of
linear Pfaffian systems
∇-de Rham cohomology theorem
Stokes phenomena and Riemann-
Hilbert-Birkhoff problem

Remark on the Notation.

A. A reference to Theorem II.4.3 is to Theorem 4.3 of Section 4 of Part II; if the part number is omitted, it is to a theorem in the part at hand.

B. Basic Notation.

1) $\mathbb{N} = \{0,1,2,\ldots\}$: set of non-negative integers.
2) $\mathbb{Z} = \{\ldots,-1,0,1,\ldots\}$: set of all integers.
3) $\mathbb{Q}$: set of all rational numbers.
4) $\mathbb{R}^+$: set of all positive real numbers.
5) $\mathbb{R}$: set of all real numbers.
6) $\mathbb{C}$: set of all complex numbers.
7) For two integers n, n', $n \geq n'$, $[n',n] = \{i \in \mathbb{N} : n' \leq i \leq n\}$.
8) For a subset J of $[1,n]$, $J^c = [1,n] - J$, the complement of J in $[1,n]$, frequently denoted by I.
9) For a subset J of $[1,n]$, $\#J$ denotes the number of elements in J.
10) For a subset J of $[1,n]$, $\mathbb{N}^J = \{(p_j)_{j\in J} : p_j \in \mathbb{N},\ j \in J\}$. An element $(p_j)_{j\in J}$ denoted by p_J. In particular, 0_J denotes $(0)_{j\in J}$.
11) For two subsets J and J' of $[1,n]$, $q_{J\cap J'}$ and $q_{J\cup J'}$ denote $(q_j)_{j\in J\cap J'}$ and $(q_j)_{j\in J\cup J'}$, respectively.
12) $x = (x_1,\ldots,x_n)$, $x_J = (x_j)_{j\in J}$ for a subset J of $[1,n]$, $x_J^{q_J} = \Pi_{j\in J} x_j^{q_j}$ for $q_J \in \mathbb{N}^J$.
13) For two elements p_J, $q_J \in \mathbb{N}^J$, $p_J \geq q_J \Longleftrightarrow p_j \geq q_j$ for all $j \in J$. $p_J \not\geq q_J \Longleftrightarrow p_j < q_j$ for some $j \in J$.

Part 0

GENERAL INTRODUCTION

An influential work of P. Deligne [11] inspired the study of differential equations with singular points (Malgrange [55], Gérard-Levelt [16], Levelt [40],etc). Moreover, it was spurred by the appearance of the sheaf of asymptotically developable functions of one variable due essentially to Y. Sibuya [73] and definitively to B. Malgrange [56] (Ramis [65,66] Martinet-Ramis [60], etc.). In the latter 1970's, we commenced to study integrable connections, i.e., locally completely integrable Pfaffian systems, in other words, systems of partial differential equations of the first order, with irregular singular points. The guiding principle was on one side the theory of ordinary differential equations with irregular singular points developed by Fabry, Poincaré, Horn, Birkhoff, Trjitzinsky, Hukuhara, Malmquist, Turrittin, Sibuya, Malgrange, Balser-Jurkat-Lutz, etc. On the other side, it was the theory of regular singularity from the perspective of algebraic geometry developed by Nilsson, Manin, Katz, Deligne, etc.

As complex analysis needs the theory of functions, the study of irregular singular points requires a theory of "asymptotic expansions" of functions.

In the one-variable case, H. Poincaré [64] introduced a definition: for a function $f(x)$ holomorphic in an open sector at the origin in $\mathbb{C}$,

$$S = S(\theta_-,\theta_+,r) = \{x\in\mathbb{C};\ \theta_- < \arg x < \theta_+,\ \ 0 < |x| < r\}\quad,$$

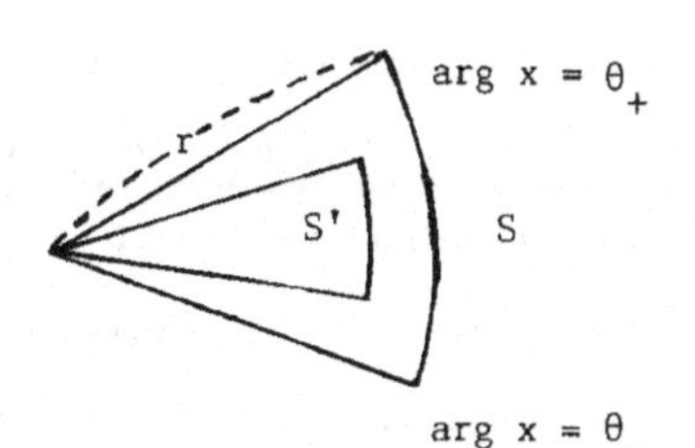

$f(x)$ is said to be <u>asymptotically developable</u> as x tends to 0 in S if there exists a formal power series $\hat{f}(x) = \sum_{i=0}^{\infty} f_i x^i$ such that for any nonnegative integer $N\in\mathbb{N}$ and for any closed subsector S' of S,

$$\sup_{x \in S'} \left| x^{-N} \left(f(x) - \sum_{i=0}^{N-1} f_i x^i \right) \right| < +\infty .$$

This concept of asymptotic expansions of functions gave the "raison d'être" to divergent power series, so that the irregular singularity of ordinary differential equations has been studied deeply since then (Birkhoff [4,5], Trjitzinsky [80], Hukuhara [27, 28], Malmquist [58], Turrittin [82], Sibuya [69], etc.).

Moreover, the concept of uniform asymptotic expansions of functions with parameters has been used in the study of ordinary differential equations with parameters and with singularities and has been developed by a great many mathematicians (cf. Wasow [83], see the enormous references). Let $f(x,y)$ be a function holomorphic in the product $S \times T$ of the open sector S and a complex manifold T. We say that $f(x,y)$ is asymptotically developable uniformly with respect to y if there exists a formal power series $\hat{f}(x,y) = \sum_{i=0}^{\infty} f_i(y)x^i$ in x with coefficients of holomorphic functions in T, such that for any nonnegative integer $N \in \mathbb{N}$ and for any closed subsector S' of S,

$$\sup_{(x,y) \in S' \times T} \left| x^{-N} \left(f(x,y) - \sum_{i=0}^{N-1} f_i(y)x^i \right) \right| < +\infty .$$

In the middle 1970's, the theory of (uniform) asymptotic expansions of functions of one variable was provided with the sheaf of functions asymptotically developable and some cohomological methods. A sector at the origin is regarded as a set of directions toward the origin "modulo radii". As the set of all directions toward 0 is identified with $S^1 = \{e^{i\theta};\ 0 \leq \theta < 2\pi\}$, a sector is written in the form $S(c,r)$ with an open set c in S^1 and a positive number $r \in \mathbb{R}^+$. For c and r, denote by $\mathcal{A}(c,r)$ the set of all functions holomorphic and asymptotically developable as x tends to 0. For fixed c, there exists a natural restriction mapping

$$i_c^{rr'} : \mathcal{A}(c,r') \longrightarrow \mathcal{A}(c,r) ,$$

for $r' \geq r$, $r', r \in \mathbb{R}^+$, together with which $\{\mathcal{A}(c,r),\ i_c^{rr'}\}$ becomes an inductive system. Put $\mathcal{A}(c) = \text{dir.lim}_{r\to 0}\,\mathcal{A}(c,r)$. Then, together with the natural restriction mapping

$$i_{cc'}: \mathcal{A}(c') \longrightarrow \mathcal{A}(c) ,$$

for $c \subset c' \subset S^1$, $\{\mathcal{A}(c), i_{cc'}\}$ becomes a presheaf over S^1 which satisfies the sheaf conditions. We denote by $\mathcal{A}$ the associated sheaf and call it the sheaf of germs of functions <u>asymptotically developable</u> as the variable tends to the origin. In a similar way, we can define the sheaf $\mathcal{A}_0$ of germs of functions <u>asymptotically developable to the identically zero series</u>; the sheaf $GL(m,\mathcal{A})$ of germs of m-by-m invertible matricial functions which are asymptotically developable; the sheaf $GL(m,\mathcal{A})_{I_m}$ of germs of functions asymptotically developable to the unit <u>matrix</u> I_m of order m. Denote by $\mathcal{O}_0$ and $\hat{\mathcal{O}}_0$ the rings of convergent and formal power series in x, respectively, and by $GL(m,\mathcal{O}_0)$ and $GL(m,\hat{\mathcal{O}}_0)$ the noncommutative rings of invertible m-by-m matrices of entries in $\mathcal{O}_0$ and $\hat{\mathcal{O}}_0$ respectively. Then, by the Borel-Ritt theorem (cf. Wasow [83]), the following sequences are <u>exact</u>:

$$0 \longrightarrow \mathcal{A}_0 \longrightarrow \mathcal{A} \longrightarrow \hat{\underline{\mathcal{O}}}_0 \longrightarrow 0 ,$$

$$I_m \longrightarrow GL(m,\mathcal{A})_{I_m} \longrightarrow GL(m,\mathcal{A}) \longrightarrow GL(m,\hat{\underline{\mathcal{O}}}_0) \longrightarrow I_m ,$$

where $\hat{\underline{\mathcal{O}}}_0$ and $GL(m,\hat{\underline{\mathcal{O}}}_0)$ denote the constant sheaves $\hat{\mathcal{O}}_0 \times S^1$ and $GL(m,\hat{\mathcal{O}}_0) \times S^1$, respectively. From them, we deduce the long exact sequences

$$0 \rightarrow 0 \rightarrow \mathcal{O}_0 \rightarrow \hat{\mathcal{O}}_0 \rightarrow H^1(S^1,\mathcal{A}_0) \rightarrow H^1(S^1,\mathcal{A}) \rightarrow H^1(S^1,\hat{\underline{\mathcal{O}}}_0) \rightarrow \cdots$$

and

$$I_m \rightarrow I_m \rightarrow GL(m,\mathcal{O}_0) \rightarrow GL(m,\hat{\mathcal{O}}_0) \rightarrow H^1(S^1,GL(m,\mathcal{A})_{I_m})$$
$$\rightarrow H^1(S^1,GL(m,\mathcal{A})) \rightarrow H^1(S^1,GL(m,\hat{\underline{\mathcal{O}}}_0)) \rightarrow \cdots .$$

Moreover, we can assert the following:

<u>Theorem A (commutative case). The following are equivalent and all are valid.</u>

1) <u>the mapping from $H^1(S^1,\mathcal{A}_0)$ to $H^1(S^1,\mathcal{A})$ is a zero mapping,</u>

2) <u>the kernel of the mapping from $H^1(S^1,\mathcal{A})$ to $H^1(S^1,\hat{\mathcal{O}}_0)$ is the neutral element,</u>

3) $\hat{\mathcal{O}}_0/\mathcal{O}_0 \simeq H^1(S^1,\mathcal{A}_0)$.

<u>Theorem B (noncommutative case). The following are equivalent and all are valid.</u>

4) <u>the mapping $H^1(S^1,GL(m,\mathcal{A})_{I_m}$ to $H^1(S^1,GL(m,\mathcal{A}))$ is a trivial mapping,</u>

5) <u>the "kernel" of the mapping $H^1(S^1,GL(m,\mathcal{A}))$ to $H^1(S^1,GL(m,\hat{\mathcal{O}}_0)$ is the neutral element,</u>

6) $GL(m,\hat{\mathcal{O}}_0)/GL(m,\mathcal{O}_0) \simeq H^1(S^1,GL(m,\mathcal{A})_{I_m})$.

These are a kind of <u>vanishing</u> theorems, which play such a role as Oka-Cartan's theorem in complex analysis. The claims 1), 2), 4) and 5) are essentially due to Y. Sibuya [73, 74] and 3) and 6) are due to B. Malgrange [56]. They utilized Theorem B (noncommutative case) for a study of a classification problem of systems of linear homogeneous ordinary differential equations with an irregular singular point at the origin, or the Riemann-Hilbert-Birkhoff problem of local version. B. Malgrange utilized Theorem A (commutative case) to obtain an isomorphism. J.-P. Ramis proved analogues of the above theorem for Gevrey class and applied them to obtain an index theorem of linear ordinary differential operators, etc. [65, 66]. Y. Sibuya also proved the analogues to study the problem of resonance [76] (cf. Lin [41, 42]).

We intended to establish a theory of asymptotic expansions of functions of several variables which is provided with the same useful tools as of one variable. We now explain our definition of <u>asymptotic developability of functions as variables tend to a singular locus</u> H. In the case where H is a coordinate hyperplane, it almost coincides with that of uniform asymptotic developability. In the following, we restrict ourselves to explaining our idea in the two-variables case.

Let $D = D(r_1)\times D(r_2)$ be a bidisc at the origin in $\mathbf{C}^2$ with holomorphic coordinates x,y and let $H\subset D$ be the locus defined by $xy = 0$. The complement of H is the product $(D(r_1) - \{0\}) \times (D(r_2) - \{0\})$ of punctured discs. Take a function $f(x,y)$ holomorphic in an open bisector $S = S(c_1,r_1) \times S(c_2,r_2)$

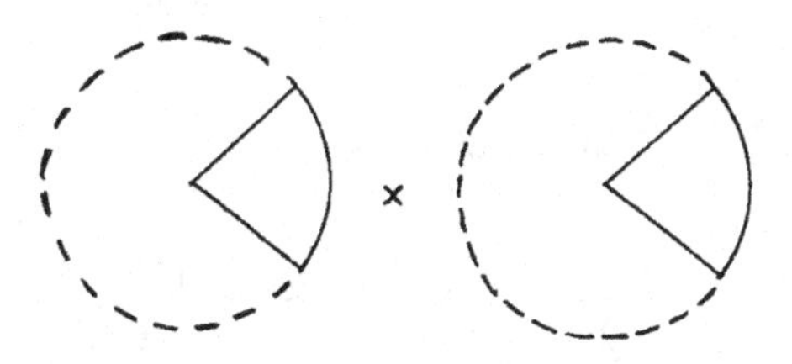

At first, let $\hat{f}(x,y) = \sum_{p=0}^{\infty}\sum_{q=0}^{\infty} f_{pq}x^p y^q$ be a formal series such that $f_{p*}(y) = \sum_{q=0}^{\infty} f_{pq}y^q$ and $f_{*q}(x) = \sum_{p=0}^{\infty} f_{pq}x^p$ are <u>convergent</u> in $D(r_2)$ and $D(r_1)$, respectively. Consider how $f(x,y)$ can be compared with $\hat{f}(x,y)$. For any nonnegative integers $M,N \in \mathbb{N}$, the sum

$$\mathrm{App}_{M,N}(\hat{f}) = \sum f_{pq}x^p y^q \qquad (p < M \quad \text{or} \quad q < N)$$

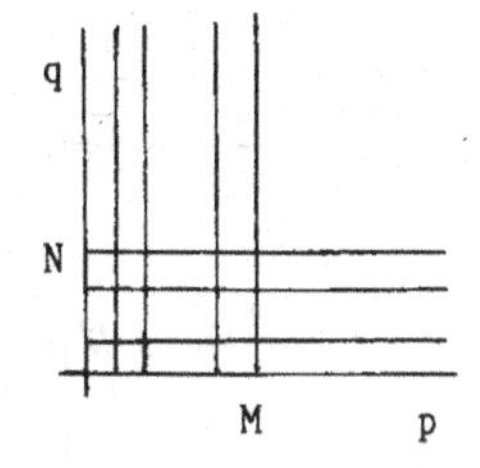

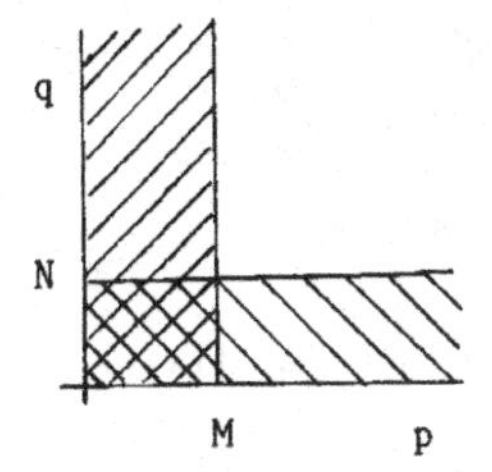

is convergent in D, especially in S. We say that $f(x,y)$ is <u>strongly asymptotically developable to</u> $\hat{f}(x,y)$ <u>as</u> (x,y) <u>tends to</u> H <u>in</u> S, if, for any $(M,N) \in \mathbb{N}^2$ and any closed subbisector S' of S,

$$\sup_{(x,y)\in S'} \left| \left(x^{-M}y^{-N}(f(x,y) - \mathrm{App}_{M,N}(\hat{f}))\right) \right| < +\infty$$

Notice that $\mathrm{App}_{M,N}(\hat{f})$ can be written in the form

$$\text{(a)}\qquad \mathrm{App}_{M,N}(\hat{f}) = \sum_{p=0}^{M-1} f_{p*}(y)x^p + \sum_{q=0}^{N-1} f_{*q}(x)y^q - \sum_{p=0}^{M-1}\sum_{q=0}^{N-1} f_{pq}x^py^q ,$$

$$\text{(b)}\qquad \mathrm{App}_{M,N}(\hat{f}) = \left(\sum_{p=0}^{M-1} f_{p*}(y) - \sum_{q=0}^{N-1} f_{pq}y^q\right) x^p$$

$$+ \left(\sum_{q=0}^{N-1} f_{*q}(x) - \sum_{p=0}^{M-1} f_{pq}x^p\right) y^q + \sum_{p=0}^{M-1}\sum_{q=0}^{N-1} f_{pq}x^py^q$$

and if $M = N$

$$\text{(c)}\qquad \mathrm{App}_{N,N}(\hat{f}) = \sum_{p=0}^{N-1}\left(\sum_{q=0}^{\infty} f_{p,p+q}y^q + \sum_{q=0}^{\infty} f_{p+q,p}x^q\right)(xy)^p .$$

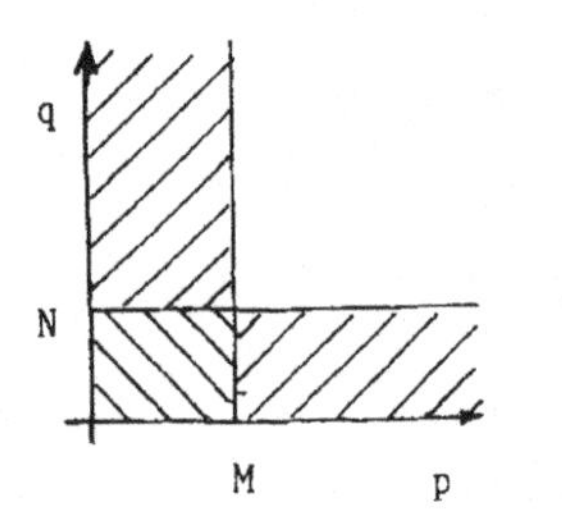

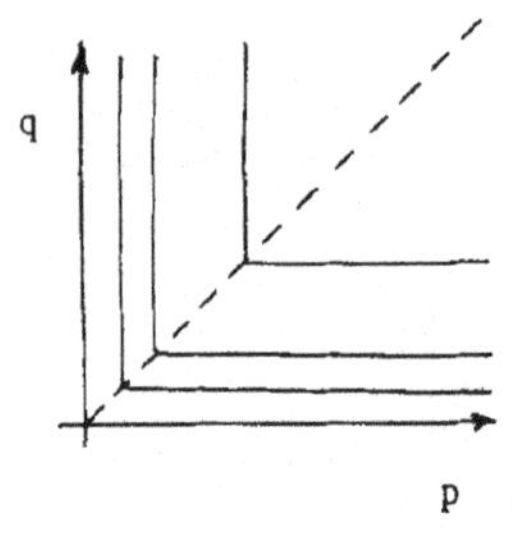

The last formula (c) implies that $\hat{f}(x,y)$ is a section of the sheaf $\mathcal{O}_{D|\hat{H}}$ over D, where

$$\mathcal{O}_{D|\hat{H}} = \text{proj.}\lim_{N\to\infty} \mathcal{O}_D/(xy)^N\mathcal{O}_D ,$$

with $\mathcal{O}_D$ the sheaf over D of germs of holomorphic functions.

In the above, we compared f(x,y) defined in S with $\mathrm{App}_{M,N}(\hat{f})$ defined in $D\supsetneq S$. So, it is possible that an interesting function defined in S is not "asymptotically developable" in the above sense. For example, the existence theorem of "asymptotic" solutions to irregular singlular Pfaffian systems is not obtained generally in the above sense. Therefore, a more general definition of "asymptotic developability" is desirable. By the first formula (a), we are led to define a concept of asymptotic developability in a wider sense. We say that $f(x,y)$ is <u>strongly asymptotically developable as</u> (x,y) <u>tends to</u> H <u>in</u> S if there exists a family

$$F = \{f_{p*}(y),\ f_{*q}(x),\ f_{pq}\}$$

of functions such that

i) $f_{p*}(y)$ and $f_{*q}(x)$ are holomorphic in $S(c_2,r_2)$ and $S(c_1,r_1)$, respectively, and f_{pq} is constant for $p,q \in \mathbb{N}$,

ii) for any $(M,N) \in \mathbb{N}^2$ and any closed subbisector S' of S,

$$\sup_{(x,y)\in S'} \left| x^{-M}y^{-N}\left(f(x,y) - App_{M,N}(F)\right)\right| < +\infty$$

where

$$App_{M,N}(F) = \sum_{p=0}^{M-1} f_{p*}(y)x^p + \sum_{q=0}^{N-1} f_{*q}(x)y^q - \sum_{p=0}^{M-1}\sum_{q=0}^{N-1} f_{pq}x^p y^q ,$$

called the approximate function of degree (M,N).

Notice that, from the definition, $f_{p*}(y)$ and $f_{*q}(x)$ are asymptotically developable to $\sum_{q=0}^{\infty} f_{pq}y^q$ and $\sum_{p=0}^{\infty} f_{pq}x^p$ in $S(c_2,r_2)$ and $S(c_1,r_1)$, respectively. We call a family F with this property a <u>consistent</u> family.

In his paper [70], Y. Sibuya utilized asymptotic expansions $\sum_{p=0}^{\infty} g_p(y)x^p$ of $f(x,y)$ such that

(1) $g_p(y)$ is holomorphic and asymptotically developable in one variable in $S(c_2,r_2)$,

(2) $f(x,y)$ is asymptotically developable to $\sum_{p=0}^{\infty} g_p(y)x^p$ uniformly in $y \in S(c_2,r_2)$,

(3) for any $N \in \mathbb{N}$, $x^{-N}\left(f(x,y) - \sum_{p=0}^{N-1} g_p(y)x^p\right)$ is holomorphic in S and asymptotically developable uniformly in $x \in S(c_1,r_1)$.

These asymptotic expansions almost coincide with ours. (We noticed this fact after our work had been completed.)

In the n variables case, the definition of strongly asymptotic developability is given in the same way except for the complexity of the definition of the approximate functions.

Under this concept of asymptotic developability, we can establish a theory of asymptotic developability together with vanishing theorems, etc. In Part I, we review the theory and in the following Parts II - IV, we show how the theory is used for completely integrable Pfaffian systems, i.e., in other words, completely integrable systems of partial differential equations with irregular singular points.

Let $S = S_1 \times \ldots \times S_{n''} \times D_{n''+1} \times \ldots \times D_n$ be a polysector at the origin in $\mathbf{C}^n$ with canonical holomorphic coordinates $x_1,\ldots,x_n$ and let U be a polydisc at the origin in $\mathbf{C}^m$ with canonical holomorphic coordinates $u_1,\ldots,u_m$. We consider in Part II integrable systems of partial differential equations with singular points on $H = \{x_1 \ldots x_{n''} = 0\}$ of the form

$$x^{p_i} e_i \frac{\partial}{\partial x_i} u = a_i(x,u) \ , \quad i = 1,\ldots,n' \leq n \ ,$$

where $p_i = (p_{i1},\ldots,p_{in''},0,\ldots,0) \in \mathbf{N}^n$, $i = 1,\ldots,n$, $e_i = x_i$ (if $i \leq n''$), $e_i = 1$ (if $i > n''$) and $a_i(x,u)$ is an m-vector function holomorphic and strongly asymptotically developable in $S \times U$. We prove existence and uniqueness theorems of strongly asymptotically developable solutions to these systems under certain general conditions. In order to construct formal power-series solutions of systems of differential equations, we provide analogues for systems of algebraic equations. By developing Hukuhara's method [28], we can prove the theorems. In the proofs, we need exact estimates of functions in polysectors. For this purpose, we introduce the notions of strictly strongly asymptotic developability and strictly consistent family in Part I. In the last of Part II, we prove splitting lemmas for completely integrable linear Pfaffian systems following Sibuya's method [69]. The existence theorems and splitting lemmas are main results of this paper by themselves on one side and preliminaries to the following Parts on the other side.

In Parts III and IV, we consider global objects, i.e., integrable connections with irregular singular points. Roughly speaking, an integrable connection is regarded as a collection of local completely integrable linear Pfaffian systems related by transformations. So, we utilize the (local) results in Part II for local studies of the integrable connections. Now let us review the concept of integrable connections.

Let M be a complex analytic manifold of dimension n and let H be a normal crossing divisor. Denote by Ω^q the sheaf of germs of holomorphic q-forms and denote by $\Omega^q(*H)$ the sheaf of germs of meromorphic q-forms with poles at most on H for $q = 0,1,\ldots,n$. We frequently use $\mathcal{O}$ and $\mathcal{O}(*H)$ instead of Ω^0 and $\Omega^0(*H)$ respectively. Let $\mathcal{S}$ be a locally free sheaf over M of $\mathcal{O}(*H)$-modules of rank m and let ∇ be an integrable connection on $\mathcal{S}$. We will view ∇ as a homomorphism of abelian sheaves

$$\nabla : \mathcal{S} \to \mathcal{S} \otimes_{\mathcal{O}(*H)} \Omega^1(*H)$$

which satisfies the Leibniz' rule

$$\nabla(f\cdot u) = u \otimes df + f \wedge \nabla u ,$$

for any local sections $f \in \mathcal{O}(*H)(U)$, $u \in \mathcal{S}(U)$ over any open set U in M, and which extends to define a structure of complex on $\mathcal{S} \otimes_{\mathcal{O}(*H)} \Omega(*H)$, the ∇-de Rham complex of $(\mathcal{S},\nabla)$

$$\mathcal{S} \xrightarrow{\nabla} \mathcal{S} \otimes_{\mathcal{O}(*H)} \Omega^1(*H) \xrightarrow{\nabla} \cdots \xrightarrow{\nabla} \mathcal{S} \otimes_{\mathcal{O}(*H)} \Omega^n(*H) \to 0 .$$

For any point $p \in M$, there exists and open set U in M containing p and a free basis $e_U = \langle e_{1U},\ldots,e_{mU}\rangle$ of $\mathcal{S}$ over U. With respect to the free basis e_U, the connection ∇ is represented by $d + \Omega_{e_U}$, i.e.,

$$\nabla(\langle e_{1U},\ldots,e_{mU}\rangle\, u) = \langle e_{1U},\ldots,e_{mU}\rangle\, (du + \Omega_{e_U} u) ,$$

where Ω_{e_U} is an m - by - m matrix of meromorphic 1-forms with poles at most on H and u is any m-vector of functions in $\mathcal{O}(*H)(U)$. By integrability

$$d\Omega_{e_U} + \Omega_{e_U} \wedge \Omega_{e_U} = 0 .$$

We call the matrix Ω_{e_U} the connection matrix with respect to e_U over U of ∇. If $f_U = \langle f_{1U},\ldots,f_{mU}\rangle$ is another free basis of $\mathcal{S}$ over U, then there exists an m-by-m matrix G of functions in $\mathcal{O}(*H)(U)$ such that

$$\langle f_{1U},\ldots,f_{mU}\rangle = \langle e_{1U},\ldots,e_{mU}\rangle G ,$$

$$\nabla(\langle f_{1U},\ldots,f_{mU}\rangle u) = \langle f_{1U},\ldots,f_{mU}\rangle(du + G^{-1}\{\Omega_{e_U} G + dG\}u) .$$

Let $x_1,\ldots,x_n$ be holomorphic local coordinates at p on U, $U \cap H = \{x_1 \ldots x_{n''} = 0\}$, then Ω_{e_U} is written in the form

$$\Omega_{e_U} = \sum_{i=1}^{n} x^{-p_i} e_i^{-1} A_i(x) dx_i ,$$

where $p_i = (p_{i1},\ldots,p_{in''},0,\ldots,0) \in \mathbb{N}^n$, $i = 1,\ldots,n$, $e_i = x_i$ (if $i \leq n''$), $e_i = 1$ (if $i \geq n''$) and $A_i(x)$ is an m-by-m matrix of holomorphic functions in U. Moreover, by integrability, we have

$$e_i \frac{\partial}{\partial x_i}\left(x^{-p_j} A_j(x)\right) + \left(x^{-p_j} A_j(x)\right)\left(x^{-p_i} A_i(x)\right)$$

$$= e_j \frac{\partial}{\partial x_j}\left(x^{-p_i} A_i(x)\right) + \left(x^{-p_i} A_i(x)\right)\left(x^{-p_j} A_j(x)\right)$$

for $i,j = 1,\ldots,n$.

In Part III, we study the kernel sheaf of ∇, i.e., the 0-th derived cohomology sheaf of the ∇-de Rham complex and in Part IV we study (partially) the higher derived cohomologies of the ∇-de Rham complex and associated complexes.

Under certain general conditions, by using the results in Part II, we investigate the structure of local fundamental matrices of solutions of the equation $\nabla u = 0$ and the relations of the matrices. By them, we describe the "Stokes phenomenon" of the equation. We regard it as a <u>locally free sheaf over the real blow-up</u> $M^{\sim}$ <u>of</u> M <u>along</u> H (see Part I) with special properties. Coversely, given a

"Stokes phenomenon", we ask whether we can construct an integrable connection with singularities which has the given Stokes phenomenon. This is the Riemann-Hilbert-Birkhoff problem of weak sense. Moreover, we ask whether it is possible to construct a global completely integrable linear Pfaffian system for the given Stokes phenomenon. This is the original Riemann-Hilbert-Birkhoff problem. We can solve the problem of weak sense by using the vanishing theorem of noncommutative case stated in Part I. By using this solution and the vanishing theorems, admitting apparent singularities, we can prove the existence of global completely integrable linear Pfaffian system for the given Stokes phenomenon on Stein manifolds or projective manifolds. Thus, the Riemann-Hilbert-Birkhoff problem is solved. These are done in Part III.

Let M^- be the real blow-up of M along H together with the natural projection $pr:M^- \to M$. We consider the following complexes of sheaves associated to the ∇-de Rham complex

(A) $$\mathcal{A}^- \otimes_{pr^*\mathcal{O}} \mathcal{S}^- \xrightarrow{\nabla^-} \mathcal{A}^- \otimes_{pr^*\mathcal{O}} (\mathcal{S}\Omega^1)^- \xrightarrow{\nabla^-} \cdots \xrightarrow{\nabla^-} \mathcal{A}^- \otimes_{pr^*\mathcal{O}} (\mathcal{S}\Omega^n)^- \to 0\ ,$$

(B) $$\mathcal{A}'^- \otimes_{pr^*\mathcal{O}} \mathcal{S}^- \xrightarrow{\nabla'^-} \mathcal{A}'^- \otimes_{pr^*\mathcal{O}} (\mathcal{S}\Omega^1)^- \xrightarrow{\nabla'^-} \cdots \xrightarrow{\nabla'^-} \mathcal{A}'^- \otimes_{pr^*\mathcal{O}} (\mathcal{S}\Omega^n)^- \to 0\ ,$$

(C) $$\mathcal{A}_0^- \otimes_{\mathbb{C}} \mathcal{S}^- \xrightarrow{\nabla_0^-} \mathcal{A}_0^- \otimes_{\mathbb{C}} (\mathcal{S}\Omega^1)^- \xrightarrow{\nabla_0^-} \cdots \xrightarrow{\nabla_0^-} \mathcal{A}_0^- \otimes_{\mathbb{C}} (\mathcal{S}\Omega^n)^- \to 0\ ,$$

where $\mathcal{S}^- = pr^*\mathcal{S}$, $(\mathcal{S}\Omega^q)^- = pr^*(\mathcal{S}\otimes_{\mathcal{O}(*H)}\Omega^q(*H))$, $\mathcal{A}^-$ is the sheaf of germs of functions strongly asymptotically developable, $\mathcal{A}'^-$ and $\mathcal{A}_0^-$ are the sheaves of germs of functions strongly asymptotically developable to $\mathcal{O}_{M|\hat{H}}$ and to 0, respectively (see Part I). We prove that "Poincaré's lemma holds" for these complexes under certain general conditions. This result is deduced from existence theorems of asymptotic solutions to integrable systems of partial differential equations established in Part II. Moreover, we consider the complexes of germ level

(D) $$(\mathcal{S})_p \xrightarrow{\nabla} (\mathcal{S}\Omega^1)_p \xrightarrow{\nabla} \cdots \xrightarrow{\nabla} (\mathcal{S}\Omega^n)_p \longrightarrow 0 ,$$

for all $p \in H$ and the complex of global section level

(E) $$\Gamma(M,\mathcal{S}) \xrightarrow{\nabla} (M,\mathcal{S}\Omega^1) \xrightarrow{\nabla} \cdots \xrightarrow{\nabla} (M,\mathcal{S}\Omega^n) \longrightarrow 0 ,$$

where $\mathcal{S}\Omega^q = \mathcal{S}\otimes_{\mathcal{O}(*H)}\Omega^q$. Notice that these complexes are also induced by taking global sections from the complexes of sheaves

(F) $$\mathcal{A}^- \otimes_{pr^*\mathcal{O}} \mathcal{S}^- \Big|_{p^-} \xrightarrow{\nabla^-} \mathcal{A}^- \otimes_{pr^*\mathcal{O}} (\mathcal{S}\Omega^1)^- \Big|_{p^-} \xrightarrow{\nabla^-} \cdots \xrightarrow{\nabla^-} \mathcal{A}^- \otimes_{pr^*\mathcal{O}} (\mathcal{S}\Omega^n)^- \Big|_{p^-} \longrightarrow 0 ,$$

for $p \in H$, $p^- = pr^{-1}(p)$ and the complex (A), respectively. We prove that <u>the first cohomology groups</u>

$$H^1((\mathcal{S}\Omega)_p, \nabla) \qquad \text{and} \qquad H^1(\Gamma(M,\mathcal{S}\Omega),\nabla)$$

<u>are isomorphic to the first cohomology groups of the kernel sheaves of</u>

$$\nabla_0^- \Big|_{p^-} : \mathcal{A}_0^- \otimes_{\mathbb{C}} \mathcal{S}^- \Big|_{p^-} \longrightarrow \mathcal{A}_0^- \otimes_{\mathbb{C}} (\mathcal{S}\Omega^1)^- \Big|_{p^-} ,$$

and

$$\nabla_0^- : \mathcal{A}_0^- \otimes_{\mathbb{C}} \mathcal{S}^- \longrightarrow \mathcal{A}_0^- \otimes_{\mathbb{C}} (\mathcal{S}\Omega^1)^- ,$$

under certain general conditions, respectively. The keys to the proof are the ∇-Poincaré's lemma and the vanishing theorem of commutative case stated in Part I. This theorem is considered as an analogue of the de Rham cohomology theorem (see section IV.1). We have not yet obtained results on the higher cohomologies ($q > 2$), because $H^q(M^-, \mathcal{A}_0^-)$ and $H^q(p^-, \mathcal{A}_0^1|_{p^-})$ may have in general complicated structures. (If $H^q(M^-, \mathcal{A}_0^-) = 0$ and $H^q(p^-, \mathcal{A}_0^-|_{p^-}) = 0$ ($q \geq 2$), then the isomorphisms are valid for the higher cohomologies.) This is a problem to be solved in the future. (See [86]. This problem is solved.)

Part I

ASYMPTOTIC DEVELOPABILITY AND VANISHING THEOREMS IN ASYMPTOTIC ANALYSIS

SECTION I.1. INTRODUCTION.

The concept of asymptotic expansions of functions of one variable was introduced by H. Poincaré in order to give an analytic meaning to formal solutions of ordinary differential equations at irregular singular points (see also section II.1). Let $f(x)$ be a holomorphic function in an open sector S at the origin in $\mathbb{C}$:

$$S = \{x \in \mathbb{C} : 0 < |x| < r, \ \theta_- < \arg x < \theta_+\} .$$

According to Poincaré's definition, $f(x)$ is said to be asymptotically developable as x tends to 0 in S if there exists a formal power series $\hat{f}(x) = \sum_{i=0}^{\infty} f_i x^i$ such that for any nonnegative integer N and for any closed subsector S' of S,

$$\sup_{x \in S'} \left| x^{-N} \left(f(x) - \sum_{q=0}^{N-1} f_q x^q \right) \right| < \infty .$$

Then, the formal power series $\hat{f}(x)$ is uniquely determined for a given function $f(x)$ and is called the asymptotic expansion of $f(x)$. Since then, the asymptotic expansions of functions have been important tools for studying the singular points of ordinary differential equations.

Poincaré's point of view is a kind of "micro-local analysis": he regarded the analysis at a point (local object) as analysis at all directions (micro-local objects) toward the point. From this point of view, in the middle 1970's, the sheaf of germs of functions asymptotically developable was introduced essentially by Y. Sibuya [73, 74] and after him definitively by B. Malgrange [56]. Using the polar coordinates, $\mathbb{C} - \{0\}$ is identified with $S^1 \times \mathbb{R}^+$ and the space of all directions toward 0 is identified with S^1. For an open set c of S^1 and a radius $r \in \mathbb{R}^+$, put

$$S(c,r) = \{x\in\mathbb{C};\ \arg x \in c,\ 0 < |x| < r\} \quad .$$

Denote by $\mathcal{A}(c,r)$ the set of all functions holomorphic and asymptotically developable in $S(c,r)$. For fixed c, there exists a natural restriction

$$i_c^{rr'}:\ \mathcal{A}(c,r') \longrightarrow \mathcal{A}(c,r)$$

for $r' \geq r$, $r',r\in\mathbb{R}^+$. Then, $\{\mathcal{A}(c,r),i_c^{rr'}\}$ is an inductive system. Put $\mathcal{A}(c) = \text{dir.}\lim_{r\to 0} \mathcal{A}(c,r)$. Then, there exists a natural restriction

$$i_{cc'} :\ \mathcal{A}(c')\longrightarrow\mathcal{A}(c)$$

for two open sets c, c' of S^1, $c\subset c'$. The set $\mathcal{A}(c)$ has naturally an algebraic structure and $\{\mathcal{A}(c),i_{cc'}\}$ becomes a presheaf over S^1 which satisfies the sheaf conditions. We denote by $\mathcal{A}$ the associated sheaf and call it the sheaf of germs of functions asymptotically developable as the variable tends to the origin. Notice that a global section of $\mathcal{A}$ over S^1 is a germ of function holomorphic at the origin, i.e.,

$$H^0(S^1,\mathcal{A}) = \Gamma(S^1,\mathcal{A}) = (\mathcal{O})_0 \ .$$

This is a reason why the asymptotic analysis could be regarded as a kind of "micro-local analysis".

Denote by $\mathcal{A}_0(c,r)$ the set of all function holomorphic and asymptotically developable to the formal power series 0 as the variable tends to the origin. By the same argument, we can construct a sheaf over S^1 from $\{\mathcal{A}_0(c,r)\}$. We denote it by $\mathcal{A}_0$ and call it the sheaf of germs of functions asymptotically developable to 0. Moreover, from the sets $GL(m,\mathcal{A}(c,r))$ of all invertible m - by - m matrices of functions in $\mathcal{A}(c,r)$, we can also define a sheaf and denote it by $GL(m,\mathcal{A})$. We denote by $GL(m,\mathcal{A})_{I_m}$ the subsheaf of $GL(m,\mathcal{A})$ of germs of m - by - m invertible matricial functions which are holomorphic and asymptotically developable to the unit matrix I_m of order m.

Suggested by works of Birkhoff [5] and Cartan (cf. [19, 25]), Y. Sibuya essentially proved

Theorem A.1). *The mapping from $H^1(S^1, GL(m,\mathcal{A})_{I_m})$ to $H^1(S^1, GL(m,\mathcal{A}))$ is trivial. Moreover, the "kernel" of the mapping $H^1(S^1, GL(m,\mathcal{A}))$ to $H^1(S^1, GL(m,(\hat{\underline{\mathcal{O}}})_0))$ is the neutral element*, where $(\hat{\underline{\mathcal{O}}})_0$ denotes the constant sheaf $(\hat{\mathcal{O}})_0 \times S^1$ with $(\hat{\mathcal{O}})_0$ the ring of formal power series.

In the proof of Theorem A.1), he also essentially proved

Theorem A.2). *the mapping from $H^1(S^1, \mathcal{A}_0)$ to $H^1(S^1, \mathcal{A})$ is a zero-mapping.*

B. Malgrange proved by C^∞-argument

Theorem B. *There exist natural isomorphisms*

1) $GL(m,(\hat{\mathcal{O}})_0)/GL(m,(\mathcal{O})_0) \simeq H^1(S^1, GL(m,\mathcal{A})_{I_m})$

2) $(\hat{\mathcal{O}})_0/(\mathcal{O})_0 \simeq H^1(S^1, \mathcal{A}_0)$.

The classical Borel-Ritt theorem (cf. [72, 83]) implies that the following sequences are exact:

$$0 \to \mathcal{A}_0 \to \mathcal{A} \to (\hat{\underline{\mathcal{O}}})_0 \to 0$$

$$I_m \to GL(m,\mathcal{A})_{I_m} \to GL(m,\mathcal{A}) \to GL(m,(\hat{\underline{\mathcal{O}}})_0) \to I_m .$$

From these short exact sequences, we deduce the long exact sequences (cf. [18])

$$0 \to 0 \to (\mathcal{O})_0 \to (\hat{\mathcal{O}})_0 \to H^1(S^1, \mathcal{A}_0) \to H^1(S^1, \mathcal{A}) \to H^1(S^1, (\hat{\underline{\mathcal{O}}})_0) \to \dots$$

and

$$I_m \to I_m \to GL(m,(\mathcal{O})_0) \to GL(m,(\hat{\mathcal{O}})_0) \to H^1(S^1, GL(m,\mathcal{A})_{I_m}) \to$$
$$H^1(S^1, GL(m,\mathcal{A})) \to H^1(S^1, GL(m,(\hat{\underline{\mathcal{O}}})_0)) \to \dots \quad .$$

So, we understand the equivalence between Sibuya's theorem A and Malgrange's theorem B.

They utilized the above theorem in the noncommutative case (i.e., A.1) or B.1)) for a study of a classification problem of "germs" of systems of linear ordinary differential equations with an irregular singular point at the origin, i.e., the Riemann-Hilbert-Birkhoff problem of local version (see also section III.1). Martinet-Ramis [60] developed their method for a study of classification problems of "germs' of nonlinear ordinary differential equations.

B. Malgrange utilized theorem A.2) or B.2) to obtain an isomorphism and, for a duality theorem, introduced the sheaf $\tilde{\mathcal{A}}$ of germs of functions asymptotically developable as the variable tends to singular points, over the "real blow-up" of the Riemann sphere with centers at the singular points. In the case where the origin is a singular point, the disjoint union $\mathbb{C} - \{0\} \cup S^1 \times \{0\}$ is regarded as the real blow-up with the center at 0. Further, the sheaf $\tilde{\mathcal{A}}$ is defined so that the restriction of $S^1 \times \{0\}$ coincides with $\mathcal{A}$ and the restriction on $\mathbb{C} - \{0\}$ coincides with $\mathcal{O}|_{\mathbb{C}-\{0\}}$.

J.-P. Ramis proved an analogue of the above theorems for Gevrey class and applied them to obtain an index theorem of linear ordinary differential operators, etc. [65, 66]. Y. Sibuya also proved the analogue to study the problem of resonance [76] (see also Lin [41, 42]).

In the several-variables case, asymptotic expansions of functions were defined by M. Hukuhara. Let $f(x_1,\dots,x_n)$ be a holomorphic function in an open polysector $S = S_1 \times \dots \times S_n$ at the origin in $\mathbb{C}^n$. We say that $f(x)$ $(x = (x_1,\dots,x_n))$ is asymptotically developable as x tends to the origin if there exists a formal power series $\hat{f}(x) = \sum_{q \in \mathbb{N}^n} f_q x^q$ such that for any nonnegative integer $N \in \mathbb{N}$ and any closed subpolysector S' of S,

$$\sup_{x \in S'} \left(\sum_{i=1}^{m} |x_i|^N \right)^{-1} \left| f(x) - \sum_{q \in \mathbb{N}^n, |q| < N} f_q x^q \right| < \infty .$$

Hukuhara utilized this concept for the study of singular points of nonlinear ordinary differential equations. Takano [78] and (partially) Gérard-Sibuya [17] utilized it for the study of singular points of completely integrable linear Pfaffian systems (see also section II.1).

The author introduced a new concept of asymptotic developability of functions of several variables in 1981 [49] after some trial and error (cf. [43, 45, 46, 47]). He defined the strongly asymptotic developability of a function as variables tend to a given singular locus (see also Part 0). This part is written for three purposes:

1. To review the definition and show the fundamental properties, i.e., theorem of Borel-Ritt type, vanishing theorems (local version), etc.,
2. to prepare the notation and to establish the fundamental facts for a regorous and delicate reasoning in Part II, i.e., strictly strongly asymptotic developability, strictly consistent family, etc.,
3. to provide some material and its important properties for global consideration in Parts III and IV, i.e., real blow-up along a normal crossing divisor, sheaf on the real blow-up, vanishing theorem of global version, etc..

Our concept of strongly asymptotic developability and our results in Part II are on the prolongation of a work of Sibuya [70]. The asymptotic expansion which Sibuya utilized in the paper almost coincide with our strongly asymptotic expansions in the two-variables case with a singular locus of a union of two hyperplanes. (We noticed this fact after our work had been completed.) He also developed his method in Gérard-Sibuya [17] for a study of Pfaffian systems with singularities of a special form (see also section II.1).

In Section 2, in the case where the singular locus H is the union of n coordinate hyperplanes at the origin in $\mathbb{C}^n$, we give the definitions of strongly symptotic developability, consistent family, strictly consistent family and strictly strongly asymptotic developability. Moreover, we prove a theorem of existence of strongly asymptotically developable functions (called Theorem of Borel-Ritt strong type) and state the vanishing theorem of local version. Two lemmas used in the following Parts are also proved. For the sake of avoiding the complexity of notation, we suppose that H is as above, but these results are also valid in the case where the singular locus H is the union of $n''(<n)$ coordinate hyperplanes at the origin in $\mathbb{C}^n$ or $\mathbb{C}^{n+m}$. We give the precise definitions of strongly asymptotic developability in this case in Section 3, where we construct the real blow-up of a complex analytic manifold along a normal crossing divisor and define the sheaf over it of germs of functions strongly asymptotically developable, etc.. The vanishing theorems of local or global version are also provided there. In Section 4, we give the asymptotic d-Poincaré's lemma and explain a little the points of problems treated in the following Parts. Moreover, we provide two fundamental properties of strongly asymptotic developability in the case where the singular locus is the union of $n''(<n)$ coordinate hyperplanes.

In the following, the paper [49] is cited by [M].

SECTION I.2. REVIEW OF STRONGLY ASYMPTOTIC DEVELOPABILITY.

Let $\mathbb{R}^+$ be the set of all positive real numbers. For an element $r = (r_1,\dots,r_n) \in (\mathbb{R}^+)^n$, we denote by $D(r)$ a polysector $D(r_1)\times\dots\times D(r_n)$ at the origin in $\mathbb{C}^n$ with radius r. Let $x = (x_1,\dots,x_n)$ be holomorphic coordinates of $D(r)$ and let H be the locus $\cup_{i=1}^n\{x\in D(r);\ x_i=0\}$. For any $i = 1,\dots,n$, we denote by $\mathcal{O}(\Pi_{j\neq i}D(r_j))[[x_i]]$ the $\mathbb{C}$-algebra of formal power series of one variable x_i with coefficients in the $\mathbb{C}$-algebra $\mathcal{O}(\Pi_{j\neq i}D(r_j))$ of holomorphic functions in $\Pi_{j\neq i}D(r_j)$, and we denote by $\mathcal{O}_H'(r)$ the intersection $\cap_{i=1}^n\mathcal{O}(\Pi_{j\neq i}D(r_j))[[x_i]]$. For r, $r'\in(\mathbb{R}^+)^n$, if $r\le r'$, then there exists a natural restriction mapping $i_{rr'}$ of $\mathcal{O}_H'(r')$ into $\mathcal{O}_H'(r)$, and so $\{\mathcal{O}_H'(r),\ i_{rr'}\}$ is an inductive system. We write $\mathcal{O}_H' = \text{dir.lim.}_{r\to 0}\mathcal{O}_H'(r)$. Note that $\mathcal{O}_H'$ is independent of the coordinates chosen in $D(r)$: if $x_i' = h_i(x)$, $i = 1,\dots,n$, is another coordinate system in $D(r)$ with H again given $\cup_{i=1}^n\{x';\ x_i' = 0\}$, then the $\mathbb{C}$-algebra $\mathcal{O}_H''$ defined by the same way as above is isomorphic to $\mathcal{O}_H'$; the isomorphism is given by

$$\sum_{p\in\mathbb{N}^n} f_p x^p \longrightarrow \sum_{p\in\mathbb{N}^n} f_p x'^p \quad .$$

Let $f(x)$ be a holomorphic function in an open polysector $S(c,r) = \Pi_{i=1}^n S(c_i,r_i)$, where c_i's are open interval on $\mathbb{R}$ the set of all real numbers, and

$$S(c_i,r_i) = \{x_i\in\mathbb{C};\ \arg x_i\in c_i,\ 0<|x_i|<r_i\} ,$$

and let $f'(x) = \sum_{p\in\mathbb{N}^n} f_p x^p$ be a formal series in $\mathcal{O}_H'(r)$. We say that $f(x)$ is <u>strongly asymptotically developable to</u> $f'(x)$ <u>as</u> x <u>tends to</u> H <u>in</u> $S(c,r)$, if, for any $N\in\mathbb{N}^n$ and for any closed subpolysector $S' = \Pi_{i=1}^n S[c_i',r_i']$, there exists a positive constant $K_{N,S'}$ such that

$$(2.1) \qquad |f(x) - \mathrm{App}_N(x;f')| \le K_{N,S'}|x|^N$$

for any $x\in S'$, where for a closed interval c_i' in $\mathbb{R}$, and $r_i'\in\mathbb{R}^+$, $i = 1,\dots,n$, we use the notation

$$S[c_i',r_i'] = \{x_i \in \mathbb{C};\ \arg x_i \in c_i',\ 0<|x_i| \leq r_i'\}\ ,$$

and the function $App_N(x;f')$ is defined by

$$(2.2)\qquad App_N(x,f') = \sum_{p\in\mathbb{N}^n,\, p\not\geq N} f_p x^p\ ,$$

and we call $App_N(x,f')$ the <u>approximate function of degree</u> N <u>of</u> f.

Given a formal series $f'(x) = \sum_{p\in\mathbb{N}^n} f_p x^p \in \mathcal{O}_H'(r)$, for any non-empty subset J of [1,n] and any $p_J = (p_j)_{j\in J} \in \mathbb{N}^J$, we put $I = J^c$ and

$$(2.3)\qquad f(x_I;p_J) = \sum_{p_I\in\mathbb{N}^I} f_{p_{I\cup J}} x_I^{p_I}\ ,$$

i.e.

$$f(x_I;p_J) = \sum_{q=(q_1,\ldots,q_n)\in\mathbb{N}^n,\, q_j=p_j(j\in J)} f_q \Pi_{i\in I} x_i^{q_i}\ ,$$

$$(2.4)\qquad PS_J(f') = \sum_{p\in\mathbb{N}^n} f(x_I;p_J) x_J^{p_J}\ .$$

Then, for any two elements f', $g' \in \mathcal{O}_H'(r)$, we have

$$(2.5)\qquad PS_J(f'+g') = PS_J(f') + PS_J(g')\ ,$$

$$(2.6)\qquad PS_J(f'g') = PS_J(f')PS_J(g')$$

for any non-empty sebset J of [1,n]. Furthermore, we put

$$(2.7)\qquad PS_J(PS_{J'}(f')) = \sum_{p_J\in\mathbb{N}^J} PS_{J'}(f(x_I;p_J)) x_J^{p_J}\ ,$$

for any two disjoint non-empty subsets J, J' of [1,n]. Then, we see easily

$$(2.8)\qquad PS_J(PS_{J'}(f')) = PS_{J'}(PS_J(f')) = PS_{J\cup J'}(f')\ .$$

For a function f holomorphic in S(c,r), there exists <u>at most one</u> formal series $f' \in \mathcal{O}_H'(r)$ to which f is strongly asymptotically developable, and so if f is strongly asymptotically developable to f', we denote $PS_J(f')$ by $FA_J(f)$ for any non-empty subset J of [1,n], and denote $App_N(x;f')$ by $App_N(x;f)$ for any $N\in\mathbb{N}^n$. In the case of J=[1,n], we use frequently FA(f) instead of $FA_{[1,n]}(f)$. Note that $App_N(x;f)$ is written in the form

$$(2.9)\quad App_N(f) = \sum_{\emptyset\neq J\subset[1,n]}(-1)^{\#J+1}\sum_{j\in J}\sum_{p_j=0}^{N_j-1} f(x_I;p_J)x_J^{p_J} ,$$

by using $f(x_I;p_J)$'s.

We denote by $\mathcal{A}'(S(c,r))$ the set of all functions which are holomorphic in the open polysector S(c,r) and respectively strongly asymptotically developable to some formal power series in $\mathcal{O}_H'(r)$ there. The set $\mathcal{A}'(S(c,r))$ is closed with respect to the fundamental operations: addition, multiplication, differentiation and integration. Moreover, each fundamental operation is commutative with the operation FA_J for any non-empty subset J of [1,n].

Let X be a domain of holomorphy included in $\mathbb{C}^n$-H and including an open polysector, and let f be a holomorphic function in X. We say that f is strongly asymptotically developable to some element in $\mathcal{O}_H'$, if, for any open polysector S(c,r) included in X, f is strongly asymptotically developable to a formal series f' in $\mathcal{O}_H'(r)$. Note that this definition does not depend on the choice of holomorphic coordinates such that the singular locus is given as the union of the n coordinate hyperplanes. We identify $\mathbb{C}^n-H = (\mathbb{C}-\{0\})^n$ with $T^n\times(\mathbb{R}^+)^n$ by the canonical mapping,

$$x = (x_1,\ldots,x_n) \longrightarrow (\arg x_1,\ldots,\arg x_n,|x_1|,\ldots,|x_n|)$$

where $T^n = (\mathbb{R}/\mathrm{mod}.2\pi\mathbb{Z})^n$, $\mathbb{R}^+ = \{r\in\mathbb{R};\ r>0\}$. For any open subset c of T^n and for any element $r = (r_1,\ldots,r_n)$ in $(\mathbb{R}^+)^n$, we define the sectorial domain S(c,r) by

$$S(c,r) = \{x\in\mathbb{C}^n;\ (\arg x_1,\ldots,\arg x_n)\in c,\ 0<|x_i|<r_i,\ i=1,\ldots,n\} ,$$

and denote by $\mathcal{A}'(c,r)$ the set of all functions which are holomorphic in the sectorial domain S(c,r) and which are respectively strongly asymptotically developable to some element in $\mathcal{O}_H'$. We see easily that $\mathcal{A}'(c,r)$ has the structure of algebra over the field $\mathbb{C}$ of complex numbers. For $r, r' \in (\mathbb{R}^+)^n$ with $r \leq r'$, we denote by $i_{r'r}$ the natural restriction mapping $\mathcal{A}'(c,r')$ into $\mathcal{A}'(c,r)$. We see easily that

$$\{\mathcal{A}(c,r),\ i_{rr'};\ r,r' \in (\mathbb{R}^+)^n\}$$

is an inductive system for any fixed open subset c of T^n. We denote by $\mathcal{A}'(c)$ the direct limit of the inductive system. For any two open subsets c, c' of T^n, the restriction mapping $i_{c'c}$ of $\mathcal{A}'(c')$ into $\mathcal{A}'(c)$ is defined if $c \subset c'$. It is easy to see that $\{\mathcal{A}'(c),\ i_{cc'}\}$ becomes a presheaf which satisfies the sheaf conditions. We denote by $\mathcal{A}'$ the associated sheaf, and call it the sheaf of germs of functions strongly asymptotically developable to $\mathcal{O}_H'$ over T^n. In the same manner, from the set

$$\mathcal{A}_0(c,r) = \{f \in \mathcal{A}'(c,r);\ FA(f) = 0\} ,$$

we obtain the sheaf of germs of functions strongly asymptotically developable to 0 over T^n. Note that the $\mathbb{C}$-algebra $\mathcal{A}'(T^n)$ of all sections over T^n in the sheaf $\mathcal{A}'$ coincides with the $\mathbb{C}$-algebra $\mathcal{O}_n$ of all germs of holomorphic functions at the origin in $\mathbb{C}^n$, namely, the set of all convergent power series of n variables, and all sections over T^n in the sheaf $\mathcal{A}_0$ coincide with the germ of the function which is identically 0 at the origin in $\mathbb{C}^n$. By a theorem of Borel-Ritt type given in [M] (THEOREM 1.(2)), we see easily that the following short sequence of sheaves

$$(2.10) \qquad 0 \longrightarrow \mathcal{A}_0 \longrightarrow \mathcal{A}' \longrightarrow \underline{\mathcal{O}}_H' \longrightarrow 0$$

is exact, where $\underline{\mathcal{O}}_H'$ is the constant sheaf of the $\mathbb{C}$-algebra $\mathcal{O}_H'$ over T^n, namely, for any open set c, $\underline{\mathcal{O}}_H'(c) = \mathcal{O}_H'$. From (1.10), we obtain the long exact sequence

$$(2.11) \qquad 0 \longrightarrow 0 \longrightarrow \mathcal{O}_n \longrightarrow \mathcal{O}_H' \longrightarrow H^1(T^n, \mathcal{A}_0) \longrightarrow H^1(T^n, \mathcal{A}') \longrightarrow \cdots$$

and a theorem of Sibuya type given in [M](THEOREM 4) implies that

<u>THEOREM</u> 2.1. <u>The mapping of</u> $H^1(T^n,\mathcal{A}_0)$ <u>into</u> $H^1(T^n,\mathcal{A}')$ <u>is a zero mapping, and so we obtain an isomorphism</u>

(2.12) $\mathcal{O}_H'/\mathcal{O}_n = H^1(T^n,\mathcal{A}_0)$.

We call (2.12) the isomorphism theorem of Malgrange type.

In order to prove the theorem of Sibuya type, as we see in [M], suggested by the equality (2.9), we treat it in a category of functions "strongly asymptotically developable" in a wider sense, and we see in the following Parts that it is natural to consider Pfaffian systems in the category of functions "strongly asymptotically developable" in the generalized sense. We now review the definition. Consider a function $f(x)$ holomorphic in an open polysector $S(c,r) = \Pi_{i=1}^n S(c_i,r_i)$. We say that $f(x)$ is <u>strongly asymptotically developable as x tends to</u> H <u>in</u> $S(c,r)$ if there exists a family of functions

$$F = \{f(x_I;q_J): \emptyset\neq J\subset[1,n],\ I=J^c,\ q_J\in\mathbb{N}^J\}$$

satisfying the following properties:

(2.13) $f(x_I;q_J)$ is holomorphic in $S_I = \Pi_{i\in I}S(r_i)$ for any non-empty proper subset J and for any $q_J\in\mathbb{N}^J$, and $f(x_\emptyset;q_{[1,n]})$ is constant for any $q_{[1,n]}\in\mathbb{N}^n$,

(2.14) for any $N\in\mathbb{N}^n$ and for any closed subpolysector S' of $S(c,r)$, there exists a constant $K_{S',N}$ such that

$$|f(x) - \mathrm{App}_N(x;F)| \leq K_{S',N}|x|^N ,$$

for any $x\in S'$, where $\mathrm{App}_N(x;F)$ is defined by the right-hand side of (2.9) with the family $F = \{f(x_I;q_J)\}$.

If f is strongly asymptotically developable, the family of functions satisfying (2.13) and (2.14) is uniquely determined. Therefore, we call the family F the <u>total family of coefficients of strongly asymptotic expansion of f(x)</u> and denote it by <u>TA(f)</u>. For a non-empty subset J of [1,n], we denote $f(x_I;q_J)$ by

$TA(f)_{q_J}$ and define the formal series

$$(2.15) \qquad FA_J(f) = \sum_{q_J \in \mathbb{N}^J} TA(f)_{q_J} x_J^{q_J}$$

which is called the <u>formal series of strongly asymptotic expansion of f(x) for</u> $J \subset [1,n]$. In particular, for $J = [1,n]$, we use $FA(f)$ instead of $FA_{[1,n]}(f)$, and call it the formal series of strongly asymptotic expansion of $f(x)$. We put $App_N(x;f) = App_N(x,F)$ and call it the <u>approximate function of degree</u> N <u>of strongly asymptotic expansion for</u> $f(x)$. It is clear that if $FA(f)$ belongs to $\mathcal{O}_H{}'(r)$, FA_J and App_N coincide with those given before respectively.

Let f be a function holomorphic and strongly asymptotically developable in an open polysector $S(c,r) = \Pi_{i=1}^n S(c_i,r_i)$. Then, for any non-empty subset J and for any $q_J \in \mathbb{N}^J$, the function $TA(f)_{q_J}$ of (n-#J) variables is strongly asymptotically developable in $\Pi_{i \in J^c} S(c_i,r_i)$, and the total family of coefficients of strongly asymptotic expansion of $TA(f)_{q_J}$ coincides with

$$(2.16) \qquad \{TA(f)_{q_{J' \cup J}}\ ;\ \emptyset \neq J' \subset J^c,\ q_{J'} \in \mathbb{N}^{J'}\}\ ,$$

that is, for any non-empty subset J' of J^c and any $q_{J'} \in \mathbb{N}^{J'}$,

$$(2.17) \qquad TA(TA(f)_{q_J})_{q_{J'}} = TA(f)_{q_{J \cup J'}}\ .$$

And so, we see that

$$(2.18) \qquad TA(TA(f)_{q_J})_{q_{J'}} = TA(TA(f)_{q_{J'}})_{q_J}\ ,$$

$$(2.19) \qquad FA_{J' \cup J}(f) = \sum_{q_J \in \mathbb{N}^J} FA_{J'}(TA(f)_{q_J}) x_J^{q_J}$$

$$= \sum_{q_{J'} \in \mathbb{N}^{J'}} FA_J(TA(f)_{q_{J'}}) x_{J'}^{q_{J'}}\ .$$

Let $S(c,r) = \Pi_{i\in[1,n]}S(c_i,r_i)$ be a open polysector and given a family of functions

(2.20) $F = \{f(x_I;q_J);\ \emptyset\neq J\subset[1,n],\ I=J^c,\ q_J\in \mathbb{N}^J\}$

such that $f(x_I;q_J)$ is holomorphic in $S_I = \Pi_{i\in I}S(c_i,r_i)$ for any non-empty proper subset J and for any $q_J\in\mathbb{N}^J$, and $f(x_\emptyset;q_{[1,n]})$ is constant for any $q_{[1,n]}$. We say that the family F is a consistent family in S(c,r), if, for any $q_J\in\mathbb{N}^J$, the function $f(x_I;q_J)$ is strongly asymptotically developable in S_I and $TA(f(x_I;q_J))$ coincides with

$$\{f(x_{I\cap I'};q_{J\cup J'});\ \emptyset\neq J'\subset I=J^c,\ I'=J'^c,\ q_{J'}\in\mathbb{N}^{J'}\},$$

namely, the following is valid: for any $N_I\in\mathbb{N}^I$ and for any closed subpolysector S_I' of S_I, there exists a constant $K_{N_I,S_I'}$ such that

$$|f(x_I;q_J)-g(x_I;q_J;N_I)| \leq K_{N_I,S_I'}|x_I|^{N_I}$$

for any x_I in S_I', where $g(x_I;q_J;N_I)$ is defined by

(2.21) $g(x_I;q_J;N_I)$

$$= \Sigma_{\emptyset\neq J'\subset I}(-1)^{\#J'+1}\Sigma_{i\in J'}\sum_{q_j=0}^{N_j-1}f(x_{I\cap I'};q_{J\cup J'})x_{J'}^{q_{J'}},$$

where I' denotes the complement of J' in I. As we see it above, if a holomorphic function f is strongly asymptotically developable in S(c,r), then the family TA(f) is consistent there.

Let S be a closed (resp. an open) polysector S[c,r] (resp. S(c,r)) and let

$$F = \{f(x_I;q_J);\ \emptyset\neq J\subset[1,n],\ I=J^c,\ q_J\in\mathbb{N}^J\}$$

be a family of functions such that $f(x_I;q_J)$ is holomorphic in the interior of

$S_I = \Pi_{i\in I} S[c_i,r_i]$ and continuous in S_I, briefly say, "holomorphic" in the closed polysector S_I (resp. holomorphic in the open polysector $S_I = \Pi_{i\in I} S(c_i,r_i)$) for any non-empty proper subset J and for any $q_J \in \mathbb{N}^J$, and $f(x_\emptyset;q_{[1,n]})$ is constant for any $q_{[1,n]}$. We say that the family F is a strictly consistent, if, for any non-empty subset J of [1,n], for any $q_J \in \mathbb{N}^J$ and for any $N_I \in \mathbb{N}^I$ ($I=J^c$), there exists a constant $K_{N_I} > 0$ such that

$$(2.22) \quad |f(x_I;q_J)-g(x_I;q_J;N_I)| \leq K_{N_I}|x_I|^{N_I},$$

for any $x_I \in S_I$, where $g(x_I;q_J;N_I)$ is defined by (2.21).

For a function f holomorphic in a closed polysector S[c,r] (resp. an open polysector S(c,r)), we say that f is strictly strongly asymptotically developable as x tends to H in S[c,r] (resp. S(c,r)), if there exists a strictly consistent family

$$F = \{f(x_I;q_J);\ \emptyset\neq J\subset[1,n],\ I=J^c,\ q_J\in\mathbb{N}^J\}$$

such that, for any $N \in \mathbb{N}^n$ and for some constant K_N, the following estimate

$$(2.23) \quad |f(x)-App_N(x;F)| \leq K_N|x|^N$$

is valid for any $x \in S[c,r]$, where $App_N(x;F)$ is defined by (2.9). By the same way as above, we define the notation TA(f), $TA(f)_{q_J}$, $FA_J(f)$, $App_N(x;f)$ etc. for this function f and we see that the same equalities are valid for this case. The set of all functions holomorphic in the closed (resp. open) polysector S[c,r] (resp. S(c,r)) and strictly strongly asymptotically developable there, is closed with respect to the fundamental operations except the differentiation. Moreover, each fundamental operation is commutative with the operation FA_J for any non-empty subset J of [1,n].

Let f be a function in an open polysector S(c,r). According to the above notations, f is strongly asymptotically developable as x tends to H in S(c,r), if and only if there exists a family of functions

$$F = \{f(x_I;p_J):\ \emptyset\neq J\subset[1,n],\ I=J^c,\ p_J\in \mathbb{N}^J\}$$

such that each function $f(x_I;p_J)$ is holomorphic in S_I for any non-empty proper subset J of $[1,n]$ and any $p_J\in \mathbb{N}^J$, $f(x_\emptyset;p_{[1,n]})$ is constant for any $p_{[1,n]}\in \mathbb{N}^n$, and for any closed (resp. proper and open) subpolysector S' of S(c,r), f is strictly strongly asymptotically developable as x tends to H in S' with TA(f) = F.

We can prove the following theorem of Borel-Ritt strong type.

THEOREM 2.2. For any strictly consistent family F in S[c,r] (or S(c,r)) there exists a function f holomorphic and strictly strongly asymptotically developable in S[c,r] (or S(c,r)) and TA(f) coincides with F.

In the following proof, we treat only the case where the sector is closed. The proof is valid in the other case, if S[c,r], $S[c_i,r_i]$ and $S[c_h,r_h]$ are replaced by S(c,r), $S(c_i,r_i)$ and $S(c_h,r_h)$, respectively.

In order to prove this theorem, we use the following lemma.

LEMMA. Let n' be an integer in [1,n], and for any subset J of [1,n] with #J=n' and for any $p_J\in \mathbb{N}^J$, let $g(x_{J^c};p_J)$ be a function holomorphic and strictly strongly asymptotically developable to 0 in S_{J^c}. Then, there exists a function g holomorphic and strictly strongly asymptotically developable in S[c,r] such that for any J with #J=n' and any $p_J\in \mathbb{N}^J$,

$$TA(g)_{p_J} = g(x_{J^c};p_J)\ .$$

PROOF OF LEMMA. We can assume without loss of generality that r_i's are less than one. We take q_i's and t_i's such that

$$0 < q_i < 1\ ,$$

$$\cos(t_i - q_i\arg x_i) \leq -1/2 \quad \text{for} \quad x_i\in S[c_i,r_i]\ .$$

For this end, it is sufficient to choose q_i's and t_i's so that

$$0 < q_i < \min\{1, \pi/3(t_{+i} - t_{-i})\}$$

$$q_i t_{+i} + 2k\pi + 5\pi/6 \leq t_i \leq q_i t_{-i} + 2k\pi + 7\pi/6$$

for some integer k, where $c_i = [t_{-i}, t_{+i}]$. For any subset J with #J=n' and for any $p_J \in \mathbb{N}^J$, if $g(x_I; p_J) \neq 0$, then we define c_{p_J} by

$$c_{p_J} = \sup\{|g(x_I;p_J)| \Pi_{i\in I} |x_i|^{-|p_J|} : x_I \in S_I'\} ,$$

and put

$$d_{p_J} = \begin{cases} \min\{1, c_{p_J}^{-1}\}, & \text{if } g(x_I;p_J) \neq 0,, \\ 0, & \text{if } g(x_I;p_J) \neq 0 . \end{cases}$$

For any J with #J=n', any $p_J \in \mathbb{N}^J$ and $h \in [1,n]$, we set

$$a_{p_J,h}(x_h) = 1 - \exp(d_{p_J} x_h^{-q_h} \exp((-1)^{1/2} t_h)) .$$

Then, we have an estimate

$$|a_{p_J,h}(x_h)| \leq d_{p_J} |x_h|^{-q_h} \leq |x_h|^{-q_h}$$

for any J with #J=n', $p_J \in \mathbb{N}^J$, $h \in J$ and $x_h \in S[c_h, r_h]$. And so, by the definition, we have the inequalities

$$|g(x_I;p_J)| \Pi_{h\in H, p_h \neq 0} |x_h|^{p_h} |a_{p_J,h}(x_h)|$$

$$\leq \Pi_{h\in H, p_h\neq 0} |x_h|^{p_h - q_h} \, \Pi_{i\in I} |x_i|^{|p_J|}$$

for any J with #J=n', $p_J \in \mathbb{N}^J$, $H \subset J$ and $p_H \in \mathbb{N}^H$.

For any subset J' of [1,n] with #J'< n' and for any $p_{J'} \in \mathbb{N}^{J'}$, we define an infinite series $g(x_{I'};p_{J'})$ $(I' = J'^{c})$ by

$$g(x_{I'};p_{J'}) = \sum_{\#J=n', J'\subset J} \sum_{p_{J-J'}\in \mathbb{N}^{J-J'}}$$

$$g(x_I;p_{J'\cup J-J'}) \prod_{j\in J-J', p_j\neq 0} x_j^{p_j} a_{p_{J'\cup J-J'},j} .$$

By using the above-stated inequalities, we can verify that these infinite series are convergent.

For any $N\in\mathbb{N}^n$, the series $g(x_{I'};p_{J'})$ can be written in the form

$$\sum_{H\subset J-J'} \sum_{h\in H} \sum_{p_h=N_h}^{\infty} \sum_{i\in J-H-J'} \sum_{p_i=0}^{N_i-1}$$

$$g(x_I;p_{J'\cup J-J'}) \prod_{j\in J-J', p_j\neq 0} x_j^{p_j} a_{p_{J'\cup J-J'},j} .$$

Therefore, the infinite series $App(x;N)$ defined by

$$App(x;N) = \sum_{\emptyset\neq J'\subset [1,n]} (-1)^{\#J'+1} \sum_{j\in J'} \sum_{p_j=0}^{N_j-1} g(x_{I'};p_{J'}) x_{J'}^{p_{J'}}$$

is equal to

$$\sum_{\#J=n'} \sum_{\emptyset\neq J'\subset J} (-1)^{\#J'+1} \sum_{j\in J'} \sum_{p_j=1}^{N_j-1} \sum_{H\subset J-J'} \sum_{h\in H} \sum_{p_h=N_h}^{\infty} \sum_{i\in J-H-J'} \sum_{p_i=0}^{N_i-1}$$

$$g(x_I;p_J) x_J^{p_J} \prod_{j\in J-J', p_j\neq 0} a_{p_J,j} = \sum_{\#J=n'} \sum_{H\subset J} \sum_{h\in H} \sum_{p_H=N_h}^{\infty} \sum_{i\in J-H} \sum_{p_i=0}^{N_i-1}$$

$$\Big(\sum_{\emptyset\neq J'\subset J-H} (-1)^{\#J'+1} \prod_{j\in J-J', p_j\neq 0} a_{p_{J'\cup J-J'},j}\Big) g(x_I;p_J) x_J^{p_J} .$$

By using the formula

$$\sum_{\emptyset\subset J'\subset J-H} (-1)^{\#J'} \prod_{j\in J-H-J'} a_{p_J,j} = \prod_{j\in J-H} (a_{p_J,j} - 1) ,$$

the infinite series

$$g(x;p_\emptyset)-App(x;N)$$

$$= \sum_{\#J=n'} \sum_{H\subset J} \sum_{h\in H}(\sum_{p_h=M}^{\infty} + \sum_{p_h=N_h}^{M-1}) \sum_{i\in J-H} \sum_{p_i=0}^{N_i-1}$$

$$\Pi_{h\in H,p_h\neq 0}\, a_{p_J,h}\, x_h^{p_h}\, \Pi_{i\in J-H,p_i\neq 0}(a_{p_J,i}-1)x_i^{p_i} g(x_I;p_J) ,$$

where $M = \max\{N_i;\ i\in J^c\}$. By using the above-stated inequalities for $g(x_I;p_J)$'s ($\#J=n'$), the strongly asymptotic developability to 0 of $g(x_i;p_J)$'s in S_I and that of $a_{p_J,h}(x_h)$'s in S_h, we can choose a positive constant K_N so that

$$|g(x;p_\emptyset)-App(x;N)| \leq K_N|x|^N$$

for any $x\in S[c,r]$. Q.E.D.

PROOF OF THEOREM 2.2.

We prove the following : for any integer $n'\in[1,n]$, there exists a function $f_{n'}$ holomorphic and strictly strongly asymptotically developable in $S[c,r]$ such that $TA(f_{n'})_{p_J} = f(x_I;p_J)$ for any J with $\#J\leq n'$ and $p_J\in \mathbb{N}^J$.

We give a proof by an induction on n'. In the case of $n' = n$, the lemma implies the statement. Suppose that the statement is true for n', and we proceed to the case of the number $n'-1$. Apply the lemma to the family $\{f(x_I;p_J)-TA(f_{n'})_{p_J};\ \#J=n'-1,\ p_J\in \mathbb{N}^J\}$. Then, we obtain a function $g_{n'-1}$ such that $TA(g_{n'-1})_{p_J} = f(x_I;p_J)-TA(f_{n'})_{p_J}$ for any $J(\#J=n'-1)$ and $p_J\in \mathbb{N}^J$. Hence, the function $f_{n'-1} = f_{n'}+g_{n'-1}$ has the above-stated property. Q.E.D.

In Part II, we prove existence theorems of strongly asymptotically developable solutions to algebraic or differential equations. In the proofs, we need the following fact concerning the validity of estimates:

Lemma 2.1. Let $F = \{f(x_I;q_J)\}$ be a strictly consistent family in $S[c,r]$

and let $v(x)$ *be a holomorphic function in* $S[c,r]$. *Take* $N_0=(N_{0,1},\ldots,N_{0,n})\in\mathbb{N}^n$. *For any* $N \geq N_0$, *let be given* $r_N=(r_{N,1},\ldots,r_{N,n})\in(\mathbb{R}^+)^n$ *such that* $r_{N,i}=r_i$ *if* $N_i=N_{0,i}$ *and* $r_{N,i}<r_i$ *otherwise*. *Suppose that for any* $N\geq N_0$,

$$\sup\{|x^{-N}(v(x)-App_N(F))| \ : x\in S[c,r_N]\}$$

is bounded. *Then, for any* $N\geq N_0$,

$$\sup\{|x^{-N}(v(x)-App_N(F))| \ : x\in S[c,r]\}$$

is bounded, i.e., $v(x)$ *is strictly strongly asymptotically developable in* $S[c,r]$ *with* $TA(v)=F$.

Proof of Lemma 2.1. For $N\in\mathbb{N}^n$, $N\geq N_0$, we set

$$ZI(N) = \{i\in[1,n] \ : N_i= N_{0,i}\} .$$

By an induction on the cardinal number $\#ZI(N)$ of $ZI(N)$, we shall prove this lemma. In the case $\#ZI(N)=n$, it is evidently true. Suppose that it is true for $\#ZI(N)=n,\ldots,n-k+1$, we proceed to the case $\#ZI(N)=n-k$. We assume, without loss of generality, that $N_i=N_{0,i}$ for $i=k+1,\ldots,n$. We write $N=(N_1,N')$, and write

$$v(x)-App_N(x) = v(x)-App_{N_{0,1},N'}(F)-(App_N(F)-App_{N_{0,1},N'}(F)) .$$

As we see that

$$App_N(F)-App_{N_{0,1},N'}(F) = \sum_{q_1=N_{0,1}}^{N_1-1}(f(q_1)-App_{N'}(f(q_1)))x_1^{q_1} ,$$

by the strict consistence of F and the hypothesis of induction, we see that

$$\sup\{|v(x)-App_N(F)||x'|^{-N'}r_{N,1}^{-N_1}; \ x\in S(c',r'), \ r_{N,1}\leq|x_1|<r_1'\}$$

is finite. Hence, we deduce that

$$\sup\{|v(x)-App_N(F)||x|^{-N}; \ x\in S(c',r')\}$$

is finite. Q.E.D.

Furthermore, we give another lemma.

Lemma 2.2. *Let $f(x)$ be a strongly asymptotically developable and let $TA(f)=\{f(x_I;q_J)\}$ be the total family of coefficients of strong asymptotic expansions for $f(x)$. Suppose for an element $N\in\mathbb{N}^n$ that $f(x_I;q_J)=0$ for any subset J of $[1,n]$ and any $q_J\in\mathbb{N}^J$ with $q_J\not\geq N_J=(N_j)_{j\in J}$. Then, $x^{-N}f(x)$ is strongly asymptotically developable and the total family of coefficients of strong asymptotic expansions of $x^{-N}f(x)$ coincides with*

$$\{x_I^{-N_I}\ f(x_I;q_J+N_J)\}\ .$$

Proof of Lemma 2.2. If $n=1$, the assertion is valid. Suppose that the assertion is valid for the numbers less than n. Then, $x_I^{-N_I}\ f(x_I;q_J+N_J)$ is strongly asymptotically developable for any J and q_J. From the inequality

$$|f(x)-App_{N'+N}(f)|\leq K_{N'+N}|x|^{N'+N}\ ,$$

for $N'\in\mathbb{N}^n$, we have

$$|x^{-N}f(x)-x^{-N}App_{N'+N}(f)|\leq K_{N'+N}|x|^{N'}\ ,$$

where $x^{-N}App_{N'+N}(f)$ coincides with the approximate function of degree N' of the consistent family $\{x_I^{-N_I}f(x_I;q_J+N_J)\}$. Q.E.D.

SECTION I.3. REAL BLOW-UP OF A COMPLEX MANIFOLD ALONG A NORMAL CROSSING DIVISOR AND VANISHING THEOREMS.

Let M be a complex manifold and let H be a divisor on M. Denote by Ω^p the sheaf over M of germs of holomorphic p-forms and denote by $\Omega^p(*H)$ the sheaf over M of germs of meromorphic p-forms that are holomorphic in M-H and have poles of arbitrary finite order on H for $p=0,\ldots,n$. In case $p=0$, we use frequently $\mathcal{O}$ (resp. $\mathcal{O}(*H)$) instead of Ω^0 (resp. $\Omega^0(*H)$).

We suppose that the divisor H has normal crossings, i.e. $H=\sum_h H_h$, where the irreducible components H_h of H are smooth and meet transversal. At a point p through which n" of the H_h pass, we may choose local holomorphic coordinates $(x_1,\ldots,x_n)$ in a neighborhood $U=\Pi_{i=1}^n\{|x_i|<r_i\}$ of $p=(0,\ldots,0)$ such that the set

$$H\cap U = \{x \ ; \ x_1\ldots x_{n''}=0\}$$

is the union of coordinate hyperplanes. The complement $U-(U\cap H)$ is a punctured polycylinder $P^*(n",n)$ given by

$$\{x : |x_i|<r_i \ (i=1,\ldots,n), \ x_1\ldots x_{n''}\neq 0\}$$

$$= \Pi_{i=1}^{n''}(D(r_i)-\{0\})\times \Pi_{i=n''+1}^{n}D(r_i) ,$$

where $D(r_i)=\{|x_i|<r_i\}$. Homotopically, $p^*(n",n)$ is a product $\times^{n''}S^1$ of n" circles, where $S^1=\{z\in\mathbb{C}: |z|=1\}$.

Let S be an open polysector at p of the form

$$S(c,r) = \Pi^{n''}_{i=1}S(c_i,r_i)\times\Pi_{i=n''+1}^{n}D(r_i) ,$$

with the coordinates system x chosen as above, where we set

$$S(c_i,r_i) = \{x_i: 0<|x_i|<r_i, \ \arg x_i\in c_i\}$$

for a positive number r_i and an open interval c_i in $\mathbb{R}$, $i=1,\ldots,n"$, and let f be a holomorphic function in S. We say that f is <u>strongly asymptotically developable</u>

as x tends to H in S at p, if there exists a family of functions

$$F = \{f(x_I;q_J): \emptyset \neq J \subset [1,n''],\ I=[1,n]-J,\ q_J \in \mathbb{N}^J\}$$

satisfying the following properties:

(3.1) $f(x_I;q_J)$ is a holomorphic function of $x_I=(x_i)_{i\in I}$ in the open polysector

$$S_I = \Pi_{i\in I\cap[1,n'']} S(r_i) \times \Pi_{j=n''+1}^{n} D(r_j)$$

for any non-empty subset J of [1,n"] and for any $q_J \in \mathbb{N}^J$,

(3.2) for any $N \in \mathbb{N}^{[1,n'']}$ and for any closed subpolysector

$$S' = \Pi_{i=1}^{n''} S[c_i',r_i'] \times \Pi_{j=n''+1}^{n} D[r_j']$$

of S, there exists a constant $K_{S',N}$ such that

$$|f(x)-App_N(x;F)| \leq K_{S',N}|x_{[1,n'']}|^N ,$$

for any $x \in S'$, where we set

$$S[c_i',r_i'] = \{x_i: 0 < |x_i| \leq r_i',\ \arg x_i \in c_i'\}$$

for a positive number $r_i' < r_i$ and a closed sub-interval c_i' of c_i, i=1,...,n",

$$D[r_j] = \{x_j: |x_j| \leq r_j\}$$

for a positive real number $r_j' < r_j$, j=n"+1,...,n, and $App_N(x;F)$ is defined by

$$(3.3)\quad App_N(x;F) = \sum_{\emptyset\neq J\subset[1,n'']} (-1)^{\#J+1} \sum_{j\in J} \sum_{p_j=0}^{N_j-1} f(x_I;p_J)x_J^{p_J} .$$

If f is strongly asymptotically developable, the family of functions satisfying (3.1) and (3.2) is uniquely determined, called the total family of coefficients of strongly asymptotic expansion of f and denoted by TA(f). For a non-empty subset J of [1,n"], we denote $f(x_I;q_J)$ by $TA(f)_{q_J}$ and define the formal series

$$(3.4)\quad FA_J(f) = \sum_{q_J\in \mathbb{N}^J} TA_J(f)_{q_J} x_J^{q_J}$$

which is called the <u>formal series of strongly asymptotic expansion of f(x) for $J\subset[1,n'']$</u>. In particular, for $J=[1,n'']$, we use FA(f) instead of $FA_{[1,n'']}(f)$, and call it the formal series of strongly asymptotic expansion of f(x). We put $App_N(x;f)=App_N(x,F)$ and call it the approximate function of degree N of strongly asymptotic expansion for f(x).

We say that a family of functions

$$(3.5)\quad F = \{f(x_I,x_{[n''+1,n]};q_J);\ \emptyset\neq J\subset[1,n''],\ I=[1,n'']-J,\ q_J\in\mathbb{N}^J\}$$

is a <u>consistent</u> family in the polysector S(c,r), if F satisfies (3.1) and

(3.6) for any $N_I\in\mathbb{N}^I$ and for any closed subpolysector S_I' of S_I, there exists a constant $K_{N_I,S_I'}$ such that

$$|f(x_I,x_{[n''+1,n]};q_J)-g(x_I,x_{[n''+1,n]};q_J;N_I)| \leq K_{N_I,S_I'}|x_I|^{N_I}$$

for any x_I, in S_I', where $g(x_I,x_{[n''+1]};N_I)$ is defined by

$$(3.7)\quad g(x_I,x_{[n''+1,n]};q_J)$$

$$= \sum_{\emptyset\neq J'\subset I}(-1)^{\#J'+1}\sum_{i\in J'}\sum_{q_j=0}^{N_j-1} f(x_{I\cap I'},x_{[n''+1,n]};q_{J\cup J'})x_{J'}^{q_{J'}},$$

where I' denotes the complement of J' in [1,n'']. If a holomorphic function f is strongly asymptotically developable in S(c,r), then the family TA(f) is consistent there. We can prove the following theorem (cf. THEOREM 2.2).

<u>THEOREM 3.1</u>. <u>For any consistent family F in S(c,r) and for any proper open subpolysector S(c',r'), there exists a function f holomorphic and strongly asymptotically developable in S(c',r') and TA(f) coincides with F</u>.

Let f be strongly asymptotically developable in S(c,r) and TA(f) = $\{f(x_I,x_{[n''+1,n]};q_J)\}$. In the case that $f(x_I,x_{[n''+1,n]};q_J)$ can be extended to a holomorphic function in $\Pi_{i\in I}D(r_i)$ for any J and q_J, we say that f is strongly asymptotically developable to the formal series

$$f'(x_{[1,n'']},x_{[n''+1,n]}) = \sum_{q\in \mathbb{N}^{n''}} f(x_{[n''+1,n]},q)\,(x_{[1,n'']})^q$$

$$= \sum_{q_j\in \mathbb{N}} f(x_{[1,n]-\{j\}};q_j)(x_j)^{q_j} \quad (j\in[1,n'']) ,$$

in $\mathcal{O}_{H\cap D(r)}{}' = \bigcap_{j=1}^{n''}\mathcal{O}(\Pi_{i\neq j}D(r_i))[[x_j]]$ as x tends to H in S at p, where $\mathcal{O}(\Pi_{i\neq j}D(r_j))[[x_j]]$ is the $\mathbb{C}$-algebra of formal series of x_j with coefficients in the $\mathbb{C}$-algebra $\mathcal{O}(\Pi_{i\neq j}D(r_j))$ of holomorphic functions in $\Pi_{i\neq j}D(r_j)$. Let I_H be the nullstellen ideal of H and put $\mathcal{O}_{M\hat{|}H} = \text{proj.lim}_{k\to\infty}\mathcal{O}/(I_H)^k$. Note that $\mathcal{O}_{H\cap D(r)}{}'$ coincides with the $\mathbb{C}$-algebra $\mathcal{O}_{M\hat{|}H}(D(r))$ of all sections of $\mathcal{O}_{M\hat{|}H}$ over D(r).

Let U be an open set included in M-H and including polysectors at any point in $cl(U)\cap H$, where cl(U) denotes the closure of U in M. We say that a holomorphic function f in U is <u>strongly asymptotically developable as variables tend to H in U</u>, if, for any point $p\in cl(U)\cap H$ with holomorphic local coordinates system x and for any open polysector $S = \Pi_{i=1}^{n''}S(c_i,r_i)\times\Pi_{i=n''+1}^{n}D(r_i)$ included in U at p, f is strongly asymptotically developable as x tends to H in S at p. We denote by $\mathcal{A}(U)$ the set of all functions holomorphic in U and strongly asymptotically developable as the variables tend to H in U, and denote by $\mathcal{A}'(U)$ (or $\mathcal{A}_0(U)$) the subset of $\mathcal{A}(U)$ of functions strongly asymptotically developable to some formal series (or the identically zero series 0) as the variables tends to H in any open polysector at any point in $cl(U)\cap H$. The set A(U) has naturally a $\mathbb{C}$-algebraic structure, and $\mathcal{A}'(U)$ and $\mathcal{A}_0(U)$ are sub-$\mathbb{C}$-algebras of $\mathcal{A}(U)$: $\mathcal{A}_0(U)$ is also a sub $\mathbb{C}$-algebra of $\mathcal{A}'(U)$. With these materials, we can define sheaves over the real blow-up along H, which is constructed in a similar manner to the construction of blow-up of a complex manifold along a submanifold as follows: we start by constructing the real blow-up of a polydisc along coordinate planes. Let D(r) be an n-dimensional disc at the origin in $\mathbb{C}^n$ with holomorphic coordinates $x_1,\ldots,x_n$, and let $V\subset D(r)$ be the locus $\bigcup_{i=1}^{n''}\{x: x_i=0\}$. Let $D(r)^-$ be the real analytic subvariety of $D(r)\times(S^1)^{n''}$ defined by the relations

$$D(r)^{-} = \{(x_1,\dots,x_n,z_1,\dots,z_{n''});\ \mathrm{Im}(x_i\bar{z}_i)=0,\ \mathrm{Re}(x_i\bar{z}_i)\geq 0,\ i\in[1,n'']\}\ .$$

The projection $pr:D(r)^{-}\longrightarrow D(r)$ on the first factor is clearly an isomorphism away from V, while, for a point $x\in V$ such that $x_i=0$, $i\in I$, and $x_j\neq 0$, $j\in[1,n'']-I$, where $I\subset[1,n'']$, the inverse image of the point x is $(S^1)^{\#I}$. The real manifold $D(r)^{-}$, together with the map $pr:D(r)^{-}\longrightarrow D(r)$, is called the real blow-up of $D(r)$ along V; the inverse image $pr^{-1}(V)$ is called the set of directions of the real blow-up. Note that the real blow-up $D(r)^{-}\xrightarrow{pr}D(r)$ is independent of the coordinates chosen in $D(r)$ such that the singular locus V is given as a union of n'' coordinate hyperplanes: let $x_i'=f_i(x)$, $i=1,\dots,n$, another coordinate system in $D(r)$. If necessary, by using permutations of indices, we can suppose that V is given as $\bigcup_{i=1}^{n''}\{x':\ x_i'=0\}$, and

$$D(r)^{-\prime}=\{(x',z')\in D(r)\times(S^1)^{n''}:\ \mathrm{Im}(x_i'\bar{z}_i')=0,\ \mathrm{Re}(x_i'\bar{z}_i')\geq 0,\ i\in[1,n'']\}$$

the blow-up of $D(r)$ in this coordinate system. Then, the isomorphism

$$f^{-}:D(r)^{-}-pr(V)\longrightarrow D(r)^{-\prime}-pr'^{-1}(V)$$

given by $x\longrightarrow f(x)$ may be extended over $pr^{-1}(V)$ by sending a point (x,z) with $x_i=0$ $(i\in I)$, $x_j\neq 0$ $(j\in[1,n'']-I)$ for some $I\subset[1,n'']$, to the point $(f(x),z')$, where,

$$z_i' = f_i(x)|f_i(x)|^{-1},\ \text{for } i\notin I$$

$$z_i' = z_i(df_i(x)/dx_i)|(df_i(x)/dx_i)|^{-1},\ \text{for } i\in I\ .$$

This last observation allows us to globalize our construction. Let $\{U_a\}$ be an open covering of M such that in each U_a with holomorphic coordinates $x_{a,i}$, $i\in[1,n]$, the subset $H\cap U_a$ may be given as the locus $\bigcup_{i=1}^{n''}\{x:\ x_{a,i}=0\}$, and let $U_a^{-}\xrightarrow{pr_a}U_a$ be the real blow-up of U_a along $U_a\cap H$. We have then isomorphisms

$$pr_{ab}:pr_a^{-1}(U_a\cap U_b)\longrightarrow pr_b^{-1}(U_a\cap U_b)$$

and using them, we can patch together the local real blow-ups U_a^{-} to form a real analytic manifold $M^{-}=\bigcup_{pr_{ab}}U_a^{-}$ with a map $M^{-}\longrightarrow M$. The manifold M^{-}, together with the map $pr: M^- \to M$, is called the real blow-up of M along H. By the construction, pr is an isomorphism away from $H \subset M$ and the following lemma is valid.

Lemma 3.1. *For any open covering* $\{U_a^-\}$ *of* M^-, *there exists such a refinement* $\{V^-(p'): p' \in pr^{-1}(p),\ p \in M\}$ *of the covering* $\{U_a^-\}$ *that* $V^-(p')$ *is an open set containing* $p' \in M^-$ *and that for any* $p \in H$, $U(p) = pr(U^-(p))$ *is a sufficiently small neighborhood of* p, *where* $U^-(p) = \bigcup_{p' \in pr^{-1}(p)} V^-(p')$.

Let M^- be the real blow-up of M along H. For an open set U^- in M^-, we define $\mathcal{A}^-(U^-)$, $\mathcal{A}'^-(U^-)$ and $\mathcal{A}_0^-(U^-)$ as follows: if $pr(U^-) \cap H = \emptyset$,

$$\mathcal{A}^-(U^-) = \mathcal{A}'^-(U^-) = \mathcal{A}_0^-(U^-) = \mathcal{O}(pr(U^-)) ,$$

if $pr(U^-) \cap H \neq \emptyset$,

$$\mathcal{A}^-(U^-) = \mathcal{A}(pr(U^-)-H) ,$$

$$\mathcal{A}'^-(U^-) = \mathcal{A}'(pr(U^-)-H) ,$$

$$\mathcal{A}_0^-(U^-) = \mathcal{A}_0'(pr(U^-)-H) .$$

Then, with the natural restriction mapping

$$i_{U^-U^{-\prime}} : \mathcal{A}^-(U^{-\prime}) \to \mathcal{A}^-(U^-)$$

for any open sets U^-, $U^{-\prime}$ in M^-, $U^- \subset U^{-\prime}$, $\{\mathcal{A}^-(U^-), i_{U^-U^{-\prime}}\}$ becomes a presheaf over M^- which satisfies the sheaf conditions. We denote by $\mathcal{A}^-$ the associated sheaf on M^- and call the sheaf of germs of functions strongly asymptotically developable (as the variables tend to the normal crossing divisor H). In the same way we obtain the sheaf $\mathcal{A}'^-$ (resp. $\mathcal{A}_0^-$) over M^- of germs of functions strongly asymptotically developable to $\hat{\mathcal{O}}_{M|H}$ (resp. 0).

For any point $p \in H$ and any open polysector $S_p(c,r)$ at p, denote by $\mathcal{CF}(S_p(c,r))$ the set of all consistent family in $S_p(c,r)$, then $\mathcal{CF}(S_p(c,r))$ has a natural $\mathbb{C}$-algebraic structure. By the same construction as above, we can define the sheaf $\mathcal{CF}^-$ over M^- of germs of consistent families:

$$(\mathcal{eF})_{p'} = \text{dir.lim}_{r\to 0, p'\in\{pr(p')\}\times c}\ \mathcal{eF}^-(S_{pr(p')}(c,r)), \quad \text{if} \quad pr(p')\in H ,$$

$$(\mathcal{eF}^-)_{p'} = 0, \quad \text{if} \quad pr(p')\notin H .$$

By the definition, there exist natural inclusions i from $\mathcal{A}_0^-$ to $\mathcal{A}'^-$ and j from $\mathcal{A}'^-$ to $\mathcal{A}^-$ and there exists natural homomorphisms FA from $\mathcal{A}'^-$ to the inverse image $pr^*(\mathcal{O}_{M|\hat{H}})$ of $\mathcal{O}_{M|\hat{H}}$ and TA from $\mathcal{A}^-$ to $\mathcal{eF}$, which is obtained by taking strongly asymptotic expansion. Then, by THEOREM 3.1, the following short sequences of sheaves over M^- are exact:

$$(3.8) \qquad 0 \longrightarrow \mathcal{A}_0^- \xrightarrow{i} \mathcal{A}'^- \xrightarrow{FA} pr^*(\mathcal{O}_{M|\hat{H}}) \longrightarrow 0 ,$$

$$(3.9) \qquad 0 \longrightarrow \mathcal{A}_0^- \xrightarrow{j\circ i} \mathcal{A}^- \xrightarrow{TA} \mathcal{eF}^- \longrightarrow 0 .$$

From the exact sequences, we obtain the long exact sequences

$$0 \to H^0(p^-, \mathcal{A}_0^-|_{p^-}) \to H^0(p^-, \mathcal{A}'^-|_{p^-}) \to H^0(p^-, pr^*(\mathcal{O}_{M|\hat{H}})|_{p^-})$$
$$\to H^1(p^-, \mathcal{A}_0^-|_{p^-}) \to H^1(p^-, \mathcal{A}'^-|_{p^-}) \to H^1(p^-, pr^*(\mathcal{O}_{M|\hat{H}})|_{p^-}) \cdots$$
$$0 \to H^0(p^-, \mathcal{A}_0^-|_{p^-}) \to H^0(p^-, \mathcal{A}^-|_{p^-}) \to H^0(p^-, \mathcal{eF}^-|_{p^-})$$
$$\to H^1(p^-, \mathcal{A}_0^-|_{p^-}) \to H^1(p^-, \mathcal{A}^-|_{p^-}) \to H^1(p^-, \mathcal{eF}^-|_{p^-}) \cdots$$

for any point $p\in H$, where $p^-=pr^{-1}(p)$, and

$$0 \to H^0(M^-, \mathcal{A}_0^-) \to H^0(M^-, \mathcal{A}'^-) \to H^0(M^-, pr^*(\mathcal{O}_{M|\hat{H}}))$$
$$\to H^1(M^-, \mathcal{A}_0^-) \to H^1(M^-, \mathcal{A}'^-) \to H^1(M^-, pr^*(\mathcal{O}_{M|\hat{H}})) \cdots$$
$$0 \to H^0(M^-, \mathcal{A}_0^-) \to H^0(M^-, \mathcal{A}^-) \to H^0(M^-, \mathcal{eF}^-)$$
$$\to H^1(M^-, \mathcal{A}_0^-) \to H^1(M^-, \mathcal{A}^-) \to H^1(M^-, \mathcal{eF}^-) \cdots$$

We see easily that

$$H^0(p^-, \mathcal{A}_0^-|_{p^-}) = 0 ,$$
$$H^0(p^-, \mathcal{A}'^-|_{p^-}) = H^0(p^-, \mathcal{A}^-|_{p^-}) = (\mathcal{O})_p ,$$

$$H^0(p^-, pr^*(\mathcal{O}_{M\hat{|}H})\big|_{p^-}) = H^0(p^-, \mathcal{G}^-\big|_{p^-}) = (\mathcal{O}_{M\hat{|}H})_p ,$$

$$H^0(M^-, \mathcal{A}_0^-) = 0 ,$$

$$H^0(M^-, \mathcal{A}'^-) = H^0(M^-, \mathcal{A}^-) = H^0(M, \mathcal{O}) ,$$

and

$$H^0(M^-, pr^*(\mathcal{O}_{M\hat{|}H})) = H^0(M^-, \mathcal{G}^-) = H^0(M, \mathcal{O}_{M\hat{|}H}) .$$

Theorem 3.2. (vanishing theorem of local version) The following are equivalent and all are valid:

(i) the image of the mapping $H^1(p^-, \mathcal{A}_0^-\big|_{p^-})$ to $H^1(p^-, \mathcal{A}'^-\big|_{p^-})$ or to $H^1(p^-, \mathcal{A}^-\big|_{p^-})$ is zero,

(ii) $H^1(p^-, \mathcal{A}_0^-\big|_{p^-})$ is isomorphic to $(\mathcal{O}_{M\hat{|}H})_p/(\mathcal{O})_p$,

(iii) the kernal of the mapping from $H^1(p^-, \mathcal{A}'^-\big|_{p^-})$ (or $H^1(p^-, \mathcal{A}^-\big|_{p^-})$ to $H^1(p^-, pr^*(\mathcal{O}_{M\hat{|}H})\big|_{p^-})$ (or to $H^1(p^-, \mathcal{G}^-\big|_{p^-})$) is zero.

Proof of Theorem 3.2. From the exactness of the sequence, we deduce the equivalence among the three statements. The assertion (i) is proved in [M] and stated as Theorem 2.1 in the case where n irreducible components of H meet at the point p. Q.E.D.

Moreover, we can prove the following:

Theorem 3.3. (vanishing theorem of global version) The followings are equivalent. If $H^1(M, \mathcal{O}) = 0$, then all are valid:

(i) the image of the mapping from $H^1(M^-, \mathcal{A}_0^-)$ to $H^1(M^-, \mathcal{A}'^-)$ or to $H^1(M^-, \mathcal{A}^-)$ is zero,

(ii) $H^1(M^-, \mathcal{A}_0^-)$ is isomorphic to $H^0(M, \mathcal{O}_{M\hat{|}H})/H^0(M, \mathcal{O})$,

(iii) the kernal of the mapping from $H^1(M^-, \mathcal{A}'^-)$ (or $H^1(M^-, \mathcal{A}^-)$) to $H^1(M^-, pr^*(\mathcal{O}_{M\hat{|}H}))$ (or to $H^1(M^-, \mathcal{G}^-)$) is zero.

Proof of Theorem 3.3. Let $\{V^-(p'): p'\in pr^{-1}(p),\ p\in M\}$ be an open covering of M^- as in Lemma 3.1, and let $\{f_{V^-(p')V^-(q')}\}$ be a 1-cocycle of the covering with coefficients in $\mathcal{A}_0^-$. Suppose that each $V^-(p')$ is sufficiently small. Then, for any point $p\in H$, $\{f_{V^-(p')V^-(p'')} : pr(p')=pr(p'')=p\}$ is a 1-cocycle of the covering $\{V^-(p'): pr(p')=p\}$ of $U^-(p)$ with coefficients in $\mathcal{A}_0^-|_{U^-(p)}$, where $U^-(p) = \bigcup_{p'\in pr^{-1}(p)} V^-(p')$. Apply Theorem 3.2 to this cocycle for any point $p\in H$. Then, there exists a 0-cochain $\{f_{V^-(p')}\}$ of the covering $\{V^-(p'): p'\in pr^{-1}(p)\}$ with coefficients in $\mathcal{A}'^-|_{U^-(p)}$ such that

$$f_{V^-(p')V^-(p'')} = f_{V^-(p'')} - f_{V^-(p')} \quad \text{on } V^-(p')\cap V^-(p'') ,$$

for $pr(p')=pr(p'')=p\in H$. Put $f_{V^-(p')}=0$ for p', $pr(p')\notin H$. By the cocycle condition, we see that

$$f_{V^-(p')} + f_{V^-(p')V^-(q')} - f_{V^-(q')} = f_{V^-(p'')} + f_{V^-(p'')V^-(q'')} - f_{V^-(q'')}$$

on $V^-(p')\cap V^-(p'')\cap V^-(q')\cap V^-(q'')$ for $p=pr(p')=pr(p'')$ and $q=pr(q')=pr(q'')$. Therefore, we can define a function $g_{U^-(p)U^-(q)}$ on $U^-(p)\cap U^-(q)$ by putting

$$g_{U^-(p)U^-(q)} = f_{V^-(p')} + f_{V^-(p')V^-(q')} - f_{V^-(q')} \quad \text{on } V^-(p')\cap V^-(q') .$$

Moreover, $g_{U^-(p)U^-(q)}$ can be regarded as a function $g_{U(p)U(q)}$ on $U(p)\cap U(q)$, where $U(b)=pr(U^-(b))$ for $b=p,q$. Thus, we obtain a 1-cocycle $\{g_{U(p)U(q)}\}$ of the covering $\{U(p): p\in M\}$ of M with coefficients in $\mathcal{O}$. By $H^1(M,\mathcal{O})=0$, there exists a 0-cochain $\{g_{U(p)}\}$ such that $g_{U(p)U(q)}=g_{U(q)}-g_{U(p)}$ on $U(p)\cap U(q)$. Hence,

$$f_{V^-(p')V^-(q')} = (f_{V^-(q')}+g_{U(q)}) - (f_{V^-(p')}+g_{U(p)}) \quad \text{on } V^-(p')\cap V^-(q')$$

and $\{(f_{V^-(p')}+g_{U(p)})\}$ is a 0-cochain of the covering $\{V^-(p'): p'\in pr^{-1}(p),\ p\in M\}$

of M with coefficients in $\mathcal{A}'^-$. This implies that (i) is valid if $H^1(M,\mathcal{O})=0$.

The equivalence among (i), (ii) and (iii) is deduced from the exactness of the long sequence. Q.E.D.

Put $\mathcal{A}'^-(*H) = \mathcal{A}'^- \otimes_{\mathrm{pr}^*\mathcal{O}} \mathrm{pr}^*(\mathcal{O}(*H))$, $\mathcal{A}^-(*H) = \mathcal{A}^- \otimes_{\mathrm{pr}^*\mathcal{O}} \mathrm{pr}^*(\mathcal{O}(*H))$, $\mathcal{O}_{M\hat{|}H}(*H) = \mathcal{O}_{M\hat{|}H} \otimes_{\mathcal{O}} \mathcal{O}(*H)$ and $\mathcal{CF}^-(*H) = \mathcal{CF}^- \otimes_{\mathrm{pr}^*\mathcal{O}} \mathrm{pr}^*(\mathcal{O}(*H))$. Then, the analogous results are valid for the following sequences of sheaves over M^-:

$$(3.10)\quad 0 \to \mathcal{A}_0^- \to \mathcal{A}'^-(*H) \to \mathrm{pr}^*(\mathcal{O}_{M\hat{|}H}(*H)) \to 0 ,$$

$$(3.11)\quad 0 \to \mathcal{A}_0^- \to \mathcal{A}^-(*H) \to \mathcal{CF}^-(*H) \to 0 .$$

For $E = \mathcal{O}_{M\hat{|}H}$, $\mathcal{O}_{M\hat{|}H}(*H)$, $\mathcal{A}'^-$, $\mathcal{A}'^-(*H)$, $\mathcal{CF}^-$, $\mathcal{CF}^-(*H)$, $\mathcal{A}^-$ and $\mathcal{A}^-(*H)$, denote by $GL(m,E)$ the sheaf of germs of invertible m - by - m matrices of the entries in E. Denote by I_m the m - by - m unit matrix and denote by $GL(m,\mathcal{A}^-)_{I_m}$ the sheaf over M^- of germs of m - by - m invertible matricial functions strongly asymptotically developable to I_m.

Then, we obtain again the short exact sequences

$$(3.12)\quad I_m \to GL(m,\mathcal{A}^-)_{I_m} \to GL(m,\mathcal{A}'^-) \to \mathrm{pr}^*(GL(m,\mathcal{O}_{M\hat{|}H})) \to I_m ,$$

$$(3.13)\quad I_m \to GL(m,\mathcal{A}^-)_{I_m} \to GL(m,\mathcal{A}^-) \to GL(m,\mathcal{CF}^-) \to I_m ,$$

$$(3.14)\quad I_m \to GL(m,\mathcal{A}^-)_{I_m} \to GL(m,\mathcal{A}'^-(*H)) \to \mathrm{pr}^*(GL(m,\mathcal{O}_{M\hat{|}H}(*H)) \to I_m ,$$

$$(3.15)\quad I_m \to GL(m,\mathcal{A}^-)_{I_m} \to GL(m,\mathcal{A}^-(*H)) \to GL(m,\mathcal{CF}^-(*H)) \to I_m .$$

From these short exact sequences, we can deduce the long exact sequences (cf. Giraud [18]) and we can prove the following theorems.

<u>Theorem.3.4.</u> <u>(vanishing theorem of local version of non-commutative case)</u> <u>For any point</u> p <u>on</u> H, <u>the following statements are equivalent and all are valid:</u>

(i) <u>the image of the mapping from</u> $H^1(p^-, GL(m,\mathcal{A}^-)_{I_m}|_{p^-})$ <u>to</u>

$H^1(p^-, GL(m, \mathcal{A}'^-|_{p^-})$ or to $H^1(p^-, GL(m, \mathcal{A}^-|_{p^-})$ is the neutral element,

(ii) $H^1(p^-, GL(m, \mathcal{A}^-)_{I_m}|_{p^-})$ is isomorphic to $GL(m, (\mathcal{O}_{M\hat{|}H})_p)/GL(m, (\mathcal{O})_p)$ as a set,

(iii) the kernel of the mapping from $H^1(p^-, GL(m, \mathcal{A}'^-)|_{p^-})$ (or from $H^1(p^-, GL(m, \mathcal{A}^-)|_{p^-})$ to $H^1(p^-, pr^*(GL(m, \mathcal{O}_{M\hat{|}H}))|_{p^-})$ (or to $H^1(p^-, GL(m, \mathcal{F})|_{p^-})$ is the neutral element.

The proof of this theorem is done by the same argument as above. The assertion (i) was proven in [M].

Theorem 3.5. (vanishing theorem of global version of non-commutative case) The following statements are equivalent. If $H^1(M, GL(m, \mathcal{O}))$ is trivial, then all are valid:

(i) the image of the mapping from $H^1(M^-, GL(m, \mathcal{A}^-))$ to $H^1(M^-, GL(m, \mathcal{A}'^-))$ or to $H^1(M^-, GL(m, \mathcal{A}^-))$ is the neutral element,

(ii) $H^1(M^-, GL(m, \mathcal{A}^-)_{I_m})$ is isomorphic to $H^0(M, GL(m, \mathcal{O}_{M|H}))/H^0(M, GL(m, \mathcal{O}))$ as a set,

(iii) the kernel of the mapping from $H^1(M^-, GL(m, \mathcal{A}'^-))$ (or from $H^1(M^-, GL(m, \mathcal{A}^-))$ to $H^1(M^-, pr^*(GL(m, \mathcal{O}_{M\hat{|}H})))$ (or to $H^1(M^-, GL(m, \mathcal{F})))$ is the neutral element.

The proof of Theorem 3.5 is the same as that of Theorem 3.3 except for the difference between the additive operation and the (non-commutative) multiplicative operation.

We do not write the statements for the long exact sequences deduced from (3.10), (3.11), (3.14) and (3.15), because we can obtain them by replacing $\mathcal{A}^-$, $\mathcal{A}'^-$, 0 etc. with $\mathcal{A}^-(*H)$, $\mathcal{A}'^-(*H)$, $\mathcal{O}(*H)$ etc., respectively.

SECTION I.4. ASYMPTOTIC d-POINCARE'S LEMMA AND FURTHER PROPERTIES OF STRONGLY ASYMPTOTIC EXPANSIONS.

Let M, H, $\mathcal{O}$, Ω^q, $\mathcal{O}(*H)$, $\Omega^q(*H)$ and (M^-, pr) be as in Section I.3. We denote by d the exterior derivative on M and use the same notation for the exterior derivative on M^-. We put

$$\mathcal{A}^-\Omega^q = \mathcal{A}^- \otimes_{pr^*\mathcal{O}} pr^*\Omega^q, \qquad \mathcal{A}^-\Omega^q(*H) = \mathcal{A}^- \otimes_{pr^*\mathcal{O}} pr^*\Omega^q(*H),$$

$$\mathcal{A}'^-\Omega^q = \mathcal{A}'^- \otimes_{pr^*\mathcal{O}} pr^*\Omega^q, \quad \mathcal{A}'^-\Omega^q(*H) = \mathcal{A}'^- \otimes_{pr^*\mathcal{O}} pr^*\Omega^q(*H)$$

and $\mathcal{A}_0^-\Omega^q = \mathcal{A}_0^- \otimes_{pr^*\mathcal{O}} pr^*\Omega^q = \mathcal{A}_0^- \otimes_{pr^*\mathcal{O}(*H)} pr^*\Omega^q(*H)$.

As an analogue of Poincaré's lemma, we can prove the following:

PROPOSITION 4.1. The following sequences of sheaves are exact:

$$\mathcal{A}^- \xrightarrow{d} \mathcal{A}^-\Omega^1 \xrightarrow{d} \cdots \xrightarrow{d} \mathcal{A}^-\Omega^n \longrightarrow 0 \tag{4.1}$$

$$\mathcal{A}'^- \xrightarrow{d} \mathcal{A}'^-\Omega^1 \xrightarrow{d} \cdots \xrightarrow{d} \mathcal{A}^-\Omega^n \longrightarrow 0 \tag{4.2}$$

$$\mathcal{A}_0^- \xrightarrow{d} \mathcal{A}_0^-\Omega^1 \xrightarrow{d} \cdots \xrightarrow{d} \mathcal{A}_0^-\Omega^n \longrightarrow 0. \tag{4.3}$$

Proof of Proposition 4.1. At a point $p' \notin pr^{-1}(H)$, the stalks are

$$(\mathcal{A}^-\Omega^q)_{p'} = (\mathcal{A}'^-\Omega^q)_{p'} = (\mathcal{A}_0^-\Omega^q)_{p'} = (\Omega^q)_{pr(p')}$$

and the results follows from the holomorphic Poincaré's lemma.

At $p \in H$, we take a neighborhood U with holomorphic coordinates $x_1,\ldots,x_n$ such that $U \cap H = \{x_1 \ldots x_{n''} = 0\}$. Let S be an open polysector at p in U. What must be verified is:

(*) Let u be a closed q-form holomorphic and strongly asymptotically developable in S. Then, there exists a (q-1)-form w holomorphic and strongly asymptotically developable in S satisfying $dw = u$ there. If u is strongly asymptotically developable to $(\mathcal{O}_{M|\hat{H}})_p$ and to 0, then w can be taken as a (q-1)-form strongly asymptotically developable to $(\mathcal{O}_{M|\hat{H}})_p$ and to 0, respectively.

For this purpose, it suffices to prove the following

Lemma. Let k be a positive integer inferior or equal to n and let u be of the form

$$u = \sum_{\#I=q, I\subset[1,k]} f_I dx_I \;.$$

Then, there exists a $(q-1)$-form F holomorphic and strongly asymptotically developable in S such that $u-dF = \sum_{\#I=q, I\subset[1,k-1]} g_I dx_I$. If u is strongly asymptotically developable to $(\hat{\mathcal{O}}_{M|H})_p$ and to 0, then F is so, respectively.

Proof of Lemma. Rewrite the q-form u as follows: $u = dx_k \wedge v_1 + v_2$, where

$$v_1 = \sum_{\#I=q, k\in I} f_I dx_{I-\{k\}} \quad \text{and} \quad v_2 = \sum_{\#I=q, k\notin I} f_I{}' dx_I \;.$$

We suppose that $du=0$. Then, for $h>k$ and $I\ni k$, $(\partial/\partial x_h)f_I=0$. By integrating f_I, we find a function F_I such that $(\partial/\partial x_k)F_I=f_I$, and $(\partial/\partial x_h)F_I=0$ for $h>k$. Set $F = \sum_{\#I=q, k\in I} F_I dx_{I-\{k\}}$, and $v=u-dF$. Then, by the definition, v is written in the form: $v = \sum_{\#I=q, I\subset[1,k-1]} g_I dx_I$. Q.E.D. of Lemma.

Hence, inductively on k, we can construct a $(q-1)$-form w satisfying $dw=u$ with strongly asymptotic developability. Q.E.D. of Proposition 4.1.

Notice that the key to the proof is to solve the system of differential equations of the first order

$$(\partial/\partial x_k)F=f, \quad (\partial/\partial x_h)F=0 \quad (k<h\leq n)$$

under the condition of complete integrability

$$(\partial/\partial x_h)f=0, \quad (k<h\leq n) \;.$$

Let Ω be an m-by-m matrix of germs of meromorphic 1-forms on the whole M with poles at most in H. Put $\nabla=d-\Omega$, and suppose $\nabla^2=0$, i.e. $d\Omega=\Omega\wedge\Omega$. Then, ∇ defines a holomorphism of sheaves of $\mathcal{A}^{-}\Omega^p(*H)\otimes_{\mathbb{C}}\mathbb{C}^m$ into $\mathcal{A}^{-}\Omega^{p+1}(*H)\otimes_{\mathbb{C}}\mathbb{C}^m$ in a natural way. We propose to study the complex of sheaves

$$(4.4) \quad \mathcal{A}^{-}(*H)\otimes_{\mathbb{C}}\mathbb{C}^m \xrightarrow{\nabla} \mathcal{A}^{-}\Omega^1(*H)\otimes_{\mathbb{C}}\mathbb{C}^m \xrightarrow{\nabla} \cdots$$
$$\xrightarrow{\nabla} \mathcal{A}^{-}\Omega^n(*H)\otimes_{\mathbb{C}}\mathbb{C}^m \longrightarrow 0 .$$

For the exactness, as above, we are led to study integrable systems of partial differential equations of the first order

$$(4.5) \quad (x_i\partial/\partial x_i)u = x^{-p_i}A_i(x)u + v_i \quad (i=n',\dots,n) ,$$

in the category of functions strongly asymptotically developable, where p_i's belong to $\mathbb{N}^n$, $A_i(x)$'s are m-by-m matrices of germs of holomorphic functions at the origin in $\mathbb{C}^n$, and v_i's are column vectors of germs of functions strongly asymptotically developable in and open polysector. More generally, we consider completely integrable systems of non-linear partial differential equations of the first order

$$(4.6) \quad (x_i\partial/\partial x_i)u = x^{-p_i}a_i(x,u) \quad (i=n',\dots,n)$$

(see Part II).

Finally, we give further two properties of strongly asymptotic developability.

Proposition 4.2. Let $f(x)$ be a function holomorphic in an open polysector $S=S_1\times\dots\times S_{n''}\times D_{n''+1}\times\dots\times D_n$ at the origin in $\mathbb{C}^n$, where S_i's are 1-dimensional open sectors and D_j's are 1-dimensional open discs. By the holomorphy, for any subset K of $[n''+1,n]$, we have the Taylor's expansion of $f(x)$ with respect to $x_K=(x_k)_{k\in K}$

$$f(x) = \sum_{q_K\in\mathbb{N}^K} f_{q_K}(x_{K^c})x_K^{q_K} .$$

If $f(x)$ is strongly asymptotically developable as x tends to $\{x_1\cdots x_{n''}=0\}$, then for any q_K, $f_{q_K}(x_{K^c})$ is strongly asymptotically developable as x_{K^c} tends to $\{x_1\cdots x_{n''}=0\}$ in $S_1\times\dots\times S_{n''}\times\Pi_{k\in K}D_k$.

This proposition is easily proved by using the Cauchy's integral formula.

Proposition 4.3. *Let* $a(x,u)$ *be a holomorphic function in an open domain* $S(c_1,r_1)\times\ldots\times S(c_{n''},r_{n''})\times D(r_{n''+1})\times\ldots\times D(r_n)\times U(R)$ *at the origin in* $\mathbb{C}^{n+m}$*, where* $S(c_i,r_i)$*'s are 1-dimensional open sectors,* $D(r_j)$*'s are 1-dimensional open discs and* $U(R)$ *is m-dimensional polydisc* (r_i*'s and* R *are radii). Let* $u(x)$ *be a m-vector of functions holomorphic and strongly asymptotically developable as* x *tends to* $x_1\ldots x_{n''}$ *in* $S(r) = S(c_1,r_1)\times\ldots\times S(c_{n''},r_{n''})\times D(r_{n''+1})\times\ldots\times D(r_n)$*. If* $a(x,u)$ *is strongly asymptotically developable as* (x,u) *tends to* $\{x_1\ldots x_{n''}=0\}$ *in* $S(r)\times U(R)$ *and if* $\lim_{x\to 0}u(x)=0$*, then* $a(x,u(x))$ *is holomorphic and strongly asymptotically developable as* x *tends to* $\{x_1\ldots x_{n''}=0\}$ *in* $S(r')$*, where* $r' \leq r$*. Moreover, the total family of coefficients of strongly asymptotic expansions of* $a(x,u(x))$ *is given as follows: for any subset* J *of the set* $[1,n'']$ *and any* $q_J \in \mathbb{N}^J$,

$$(**)\qquad TA(a(x,u(x)))_{q_J} = (q_J!)^{-1}(\partial/\partial x_J)(a(x,\hat{u}(x)))\Big|_{\hat{u}(x)=FA_J(u(x)),\,x_J=0_J}$$

Proof. For any $N \in \mathbb{N}^{[1,n'']}$, let $App_N(u(x))$ be the approximate function of degree N of $u(x)$. Denote by G the family defined as in (**) and let $App_N(G)$ be the approximate function of degree N of the family G. We shall prove that

$$a(x,u(x)) - a(x,App_N(u(x))) = O(|x|^N)$$

and

$$a(x,App_N(u(x))) - App_N(G) = O(|x|^N) ,$$

from which we conclude

$$a(x,u(x)) - App_N(G) = O(|x|^N) .$$

The former estimate is deduced from

$$a(x,u(x)) - a(x,App_N(u(x))) = \int_0^1 (\partial/\partial t)a(x,tu(x)+(1-t)App_N(u(x)))dt .$$

The latter estimate is deduced from

$$a(x,App_N(u(x))) - App_N(G)$$

$$=\int_0^{x_1}\cdots\int_0^{x_{n''}}(x_1-t_1)^{N_1-1}\cdots(x_{n''}-t_{n''})^{N_{n''}-1}((N_1-1))^{-1}\cdots((N_{n''}-1))^{-1}$$

$$(\partial/\partial t)^N a(t,x_{[n''+1,n]},App_N(u(t,x_{[n''+1,n]})))dt_1\cdots dt_{n''}\ ,$$

where $(\partial/\partial t)^N=(\partial/\partial t_1)^{N_1}\cdots(\partial/\partial t_{n''})^{N_{n''}}$. Q.E.D.

Part II

EXISTENCE THEOREMS OF ASYMPTOTIC SOLUTIONS AND SPLITTING LEMMAS

SECTION II.1. INTRODUCTION

About one century ago, two French mathematicians, Fabry and Poincaré, made important contributions to the field of ordinary differential equations with irregular singular points. Consider a linear homogeneous ordinary differential equation of the m-th order in $\mathbb{C}$,

$$(1)\quad \left(\frac{d}{dx}\right)^m u + x^{-p_{m-1}} a_{m-1}(x) \left(\frac{d}{dx}\right)^{m-1} u + \dots + x^{-p_0} a_0(x)u = 0 ,$$

where the p_i's are integers and the $a_i(x)$'s are functions holomorphic at the origin with $a_i(0) \neq 0$ for $i=0,\dots,m-1$. The origin is an irregular singular point if the so-called irregularity

$$\max \{(m-i)^{-1} p_i ;\ i=0,\dots,m-1\} = \frac{r}{q} \in \mathbb{Q}$$

is greater than one. In 1885, Fabry [12] proves

<u>Theorem</u> A. <u>For the above differential equation, there exists</u> m <u>independent formal solutions of the form</u>

$$f_{ih} = \left\{ \sum_{j=0}^{h} \hat{p}_{ihj}(t)(\log t)^j \right\} t^{\mu_i} \exp(\lambda_i(t)) , \quad 0 \leq h \leq m_i - 1, \quad 1 \leq i \leq s ,$$

<u>where</u> $\sum_{i=1}^{s} m_i = m$, $t = x^{1/q}$, $\lambda_i(t)$'s <u>are polynomials in</u> t^{-1}, μ_i's <u>are constants and</u> $\hat{p}_{ihj}(t)$'s <u>are formal power series in</u> t. In general, $\hat{p}_{ihj}(t)$'s are <u>divergent</u>.

In 1886, Poincaré [64] introduced the concept of asymptotic expansions of functions of one variable and gave an <u>analytic</u> meaning to Fabry's formal solutions under a certain restriction on the differential equation. After Poincaré's work, many mathematicians were interested in the problem and tried to prove the existence of analytic solutions which are asymptotically developable to the formal solutions: Sternberg, Horn, Birkhoff, etc. (cf. Ince [29], Kohno-Okubo[39]). Finally, in 1933, Trjitzinsky [80] succeeded in proving the following:

Theorem B. For any direction ℓ toward 0, there exist an open sector S_ℓ containing ℓ and functions $p^{\ell}_{ihj}(t)$ holomorphic and asymptotically developable to $\hat{p}_{ihj}(t)$ in S_ℓ, such that the m functions

$$\left\{\sum_{j=0}^{h} p^{\ell}_{ihj}(t)(\log t)^j\right\} t^{\mu_i}\exp(\lambda_i(t)) , \qquad 0\le h\le m_i-1, \quad 1\le i\le s ,$$

form a fundamental solution system of the differential equation in S_ℓ.

We now pass to the studies of systems of linear homogeneous ordinary differential equations with an irregular singular point at the origin in $\mathbb{C}$

$$(2) \qquad x^{r+1}\frac{d}{dx}u = A(x)u ,$$

where r is a positive integer and $A(x)$ is an m - by - m matrix of holomorphic functions at the origin. Birkhoff made important contributions for a generic case [4,5] (see also Section III.1). In 1937, Hukuhara [27] proved the following.

Theorem C. There exists a fundamental matrix of formal solutions of the form

$$\hat{p}(t)t^{M}\exp(\Lambda(t)) ,$$

where $t=x^{1/q}$ for some positive integer q, $\Lambda(t)$ is a diagonal matrix of polynomials in t^{-1}, M is a constant matrix commuting with $\Lambda(t)$ (i.e., $\Lambda(t)M=M\Lambda(t)$), and $\hat{P}(t)$ is an invertible matrix of formal series in t. In other words, by the formal transformation

$$(3) \qquad u = \hat{P}(t)v ,$$

the system is transformed to the system

$$(4) \qquad \frac{d}{dt} v = \left(\frac{d}{dt} \Lambda(t) + \frac{M}{t}\right) v \quad .$$

Theorem D. For any direction ℓ toward 0, there exist an open sector S_ℓ containing ℓ and an invertible m-by-m matrix $P_\ell(t)$ holomorphic and asymptotically developable to $\hat{P}(t)$ in S_ℓ such that

$$P_\ell(t)t^M \exp(\Lambda(t))$$

forms a fundamental matrix of holomorphic solutions in S_ℓ.

In 1955, Turrittin [82] reproved Theorem C in a way different from that of Hukuhara, and so it is now called the Hukuhara-Turrittin formal reduction theorem for systems of linear differential equations at singular points.

Certain systems of nonlinear ordinary differential equations with singular point at the origin in $\mathbb{C}$,

$$(5) \qquad x^{r+1} \frac{d}{dx} w = a(x,w) \ ,$$

where r is a nonnegative integer and $a(x,w)$ is an m-vector of holomorphic functions with respect to $(x,w) \in \mathbb{C}^{m+1}$, admit formal power series solutions. For example, Euler's differential equation

$$x^2 \frac{d}{dx} w - w = x$$

has a formal power series solution $\hat{w} = \sum_{i=1}^{\infty} n!\, x^{n+1}$. In 1941, by using the formal reduction theorem, Malmquist [58] proved

Theorem E. Suppose that a system (5) of nonlinear ordinary differential equations has a formal power series solution $\hat{w}(x)$. Then, for any direction ℓ toward 0, there exist an open sector S_ℓ containing ℓ and a solution of the system which is holomorphic and asymptotically developable to the formal solution $\hat{w}(x)$ in S_ℓ.

As we mentioned before, transformation (3) reduces system (2) to system (4). This implies that the matrix $\hat{P}(t)$ is a formal power series solution of the equation

$$(6) \qquad \frac{d}{dt} P(t) = t^{-(r+1)q} A(t^q) P(t) - P(t) \left\{ \frac{d}{dt} \Lambda(t) + \frac{M}{t} \right\}.$$

Therefore, Theorem D can be regarded as a corollary of Theorem E. On the other hand, although in his paper [28] Hukuhara treated only systems of linear non-homogeneous ordinary differential equations, we know that his method can be utilized for systems of nonlinear ordinary differential equations as well. Furthermore, Hukuhara constructed the sectors more precisely than Malmquist did, and he showed very clearly how the asymptotic solutions depend on initial values (cf. Iwano [30]). Therefore, we now call Theorems D and E the Hukuhara-Malmquist <u>Theorems of existence of asymptotic solutions</u> to systems of ordinary differential equations at singular points.

In the formal reduction process for systems of homogeneous linear ordinary differential equations, genuine formal transformations are utilized only in the so-called blockdiagonalization or splitting process, i.e., the procedure in whcih the whole system of differential equations is decomposed into several blocks according to the blocks of the leading matrix of the system. More precisely, consider a system of linear homogeneous differential equations

$$(7) \qquad x^{\nu+1} \frac{d}{dx} u = B(x) u ,$$

where ν is a positive integer, $B(x)$ is an m-by-m matrix of functions holomorphic and asymptotically developable in an open sector S at the origin in $\mathbb{C}$ and $B_0 = \lim_{x \to 0} B(x)$ is written in the form

$$B_0 = \oplus_{i=1}^{s} (d_i I_{m_i} + T_i) ,$$

where the d_i's are different eigenvalues of B_0, m_i is the multiplicity of d_i, I_{m_i} is the unit matrix of order m_i and T_i is a strictly triangular constant matrix for $i=1,\ldots,s$. Then, Sibuya [69] proved the following:

<u>Theorem</u> F. <u>For any direction</u> ℓ <u>in</u> S, <u>there exist an open sector</u> S_ℓ <u>containing</u> ℓ <u>and included in</u> S, <u>and an</u> m-<u>by</u>-m <u>invertible matrix</u> $P_\ell(x)$ <u>holomorphic</u>

and asymptotically developable in S_ℓ such that by the transformation $u=P_\ell(x)v$, the original system is changed into the new block-diagonal system

$$x^{\nu+1}\frac{d}{dx}v = (\oplus_{i=1}^{s} C_i(x))v ,$$

where $C_i(x)$ is an m_i - by - m_i matrix of functions holomorphic and asymptotically developable in S_ℓ and

$$\lim_{x\to 0} C_i(x) = d_i I_{m_i} + T_i , \quad i = 1,\dots,s .$$

Sibuya's point of view is useful in many other contexts. In fact, he proved a more general theorem than the above for systems of linear homogeneous ordinary differential equations with parameters (cf. [26,70] etc.).

In the study of singular points of ordinary differential equations, it is fundamental and important to prove Theorems A - F. The purpose of this Part is to prove analogues of these results for the completely integrable systems of partial differential equations of the first order, in other words, the completely integrable Pfaffian systems.

In the study of singular points of differential equations, there is another important topic: characterization of regular singularity. In the 1960's, it became a focus of particular interest in two fields of mathematics—ordinary differential equations and algebraic geometry. It was extensively studied by Moser [61], Lutz, Jurkat-Lutz [32], Harris [23], Malgrange [56], etc. in the former field and by Nilsson [63], Manin [59], Brieskorn [6], Griffiths [20], Katz [34], Deligne [11], etc. in the latter. With regard to this problem, Deligne proved the so-called existence of cyclic vectors which allows us, locally speaking, to consider systems of linear homogeneous ordinary differential equations of the first order as linear homogeneous ordinary differential equations of higher order. (As a matter of fact, Cope [10] had stated this fact explicitly and proved Theorem C in a slightly weaker form.) Note that the characterization of regular singularity is well known for linear homogeneous ordinary differential equations of higher order (Fuchs, cf.

Ince [29], Coddington-Levinson [9]). Therefore, cyclic vectors are useful in the study of systems of differential equations of the first order. Moreover, the idea of cyclic vectors suggests a somewhat intrinsic method of constructing formal reductions. For example, after Deligne, Levelt [40] reproved Theorem C by using the existence of cyclic vectors. Baldassarri [1] translated the Hukuhara-Turrittin formal reduction theorem into a decomposition theorem of differential modules in the language introduced by Manin. Furthermore, Malgrange [57] and Robba [67] proved that a linear ordinary differential operator has a canonical formal decomposition as an operator (cf. Ince [29]) and translated the fact into a decomposition theorem of differential modules. (Recently a characterization of regular singularity was obtained by Kitagawa [38] in which there is a reduction method different from Hukuhara's and Turrittin's for systems of differential equations.)

In the mid-1970's, the sheaf of germs of asymptotically developable functions of one variable was introduced essentially by Y. Sibuya [73, 74] and after him definitively by B. Malgrange [56]; they developed treatments of ordinary differential equations using the sheaf of germs of functions having asymptotic expansions.

The study of Pfaffian systems with singularities was opened by Gérard [13] and Deligne [11]. They studied Pfaffian systems with regular singular points. The representation of a fundamental solution matrix was studied abstractly by Deligne, slightly abstractly and slightly concretely by Gérard-Levelt [15] and concretely by Takano-Yoshida [79]. After these results, the study of Pfaffian systems with irregular singular points was opened by Gérard-Sibuya [17] and Takano [78]. They treated Pfaffian systems of special form:

$$(8) \qquad \left(d - \sum_{i=1}^{n} x_i^{-p_i-1} A_i(x)dx_i\right) u = 0 ,$$

where p_i is a nonnegative integer and $A_i(x)$ is an m-by-m matrix of holomorphic functions at $0 \in \mathbb{C}^n$ for $i=1,\ldots,n$. In particular, Gérard-Sibuya proved a sensational result. Consider a completely integrable nonlinear Pfaffian equation

$$(9) \qquad du - \sum_{i=1}^{n} x_i^{-p_i-1} a_i(x,u) = 0$$

where $a_i(x,u)$ is holomorphic at $(0,0) \in \mathbb{C}^{n+m}$, p_i is a nonnegative integer, $B_i = (\partial a/\partial u)|_{x=0,u=0}$ is invertible if $p_i > 0$ or B_i has no eigenvalues of integer if $p_i = 0$ for $i=1,\ldots,n$.

<u>Theorem</u> G. <u>Any formal power series solution</u> $\hat{u}(x)$ <u>with</u> $u(0)=0$ <u>is convergent in a full neighborhood of the origin in</u> $\mathbb{C}^n$.

Suppose that the $A_i(0)$'s are simultaneously decomposed into blocks of the same size,

$$A_i(0) = \oplus_{j=1}^{s}(d_{ij} I_{m_i} + T_{ij})$$

where $d_{ij} \neq d_{ij'}$ if $p_i > 0$ and $j \neq j'$, and $d_{ij} - d_{ij'} \notin \mathbb{Z}$ if $p_i = 0$ and $j \neq j'$, T_{ij} is an m_i - by - m_i strictly upper triangular matrix for $i=1,\ldots,n$, $j=1,\ldots,s$. Then, by Theorem G, we deduce

<u>Theorem</u> H. <u>By a holomorphic transformation</u> $u=P(x)v$, <u>the above linear Pfaffian system is changed into a block-diagonal system</u>

$$\left\{ d - \sum_{i=1}^{n} x_i^{-p_i-1} (\oplus_{j=1}^{s} B_{ij}(x)) dx_i \right\} v = 0$$

<u>where</u> $B_{ij}(x)$ <u>is holomorphic at</u> 0 <u>for</u> $i=1,\ldots,n$, $j=1,\ldots,s$.

At the time, Theorem G and H seemed incredible, conflicting with common sense in the field of ordinary differential equations with singular points, so Sibuya [75] prudently reproved them in ways different from his original proof.

The author was interested in the sheaf-theoretical method developed by Y. Sibuya and B. Malgrange and commenced to generalize their results to the several-variables case. After some trial and error, he reproved Theorem H in a slightly sheaf-theoretic way under a restriction on systems (8) [43]. In his work, he adopted a slightly different definition of asymptotic developability from

Hukuhara's (see Section I.1). This was a first step to the concept of strongly asymptotic developability. Moreover, by calculating the Pfaffian systems of the confluent hypergeometric functions of two variables, he pointed out that these systems are not written in the form considered by preceding authors [44]. He started to consider completely integrable linear Pfaffian systems with singular points on normal crossing divisors of general form

$$(10) \qquad \left(d - \sum_{i=1}^{n} x_1^{-p_{i1}} \cdots x_{n''}^{-p_{in''}} e_i^{-1} A_i(x) dx_i\right) u = 0 ,$$

where $n'' \leq n$, $p_{ij} \in \mathbb{N}$, $e_i = x_i$ (if $1 \leq i \leq n''$), $e_i = 1$ (if $n'' < i \leq n$) and $A_i(x)$ is an m-by-m matrix of holomorphic functions at the orign in $\mathbb{C}^n$. After more trial and error [45-47], in 1981 the concept of strongly asymptotic developability came to him (with his future wife). By using this concept, we can prove a kind of vanishing theorems (see Part I) and existence theorems of strongly asymptotic solutions for integrable systems of partial differential equations of the first order. As we recently noticed, a work of Sibuya in 1968 [70] also suggested research in this direction.

In the following sections, in the notation of Part I, we prove existence theorems of integrable systems of partial differential equations of the first order under certain general conditions by developing Hukuhara's method [28]. In order to construct formal power series solutions of systems of differential equations, we provide analogues for systems of algebraic equations. For the sake of simplicity, we first treat (Section II.2) systems of differential equations with respect to $y = x_1$, regarding $x_2, \ldots, x_n$ as parameters. In Section II.3 we treat integrable systems

$$(11) \qquad x_1^{p_{i1}} \cdots x_{n''}^{p_{in''}} e_i \frac{\partial}{\partial x_i} u = a_i(x,u) \qquad i=1,\ldots,n' \leq n ,$$

under certain general conditions.

Finally, following Sibuya's method [69], we establish splitting lemmas for our case. These theorems are main theorems of this paper as well as preliminaries to Parts III and IV.

An analogue of Theorem C, i.e., formal reduction theorem for Pfaffian systems (8) of special forms, was proven by Charièrre-Gérard [8] (cf. Charièrre [7]) and Levelt-van den Essen [84] without using complex blow-ups. For general Pfaffian systems, complex blow-ups are indispensable for reducing the systems (10) and the author has been preparing an article on this topic (cf. Majima [48]). He also is preparing a joint paper with Y. Sibuya on characterizations of regular singularity in the several-variable case [85].

Remark to readers: in Sections 2 and 3, for the sake of avoiding the complexity of notation, we suppose that the singular locus $H=\{x_1 \ldots x_n=0\}$ in $\mathbb{C}^n$ and $H'=\{x_1 \ldots x_n=0\}$ in $\mathbb{C}^{n+m}$. The proofs of existence theorems in these sections are also valid with simple modifications in the case where the singular locus $H=\{x_1 \ldots x_{n''}=0\}$ in $\mathbb{C}^n$ and $H'=\{x_1 \ldots x_{n''}=0\}$ in $\mathbb{C}^{n+m}$ for $n''<n$. We state the existence theorems for this case (including the above case except for the corollaries) in Section without proofs. The reader can use Section 4 for a list of existence theorems cited in the last two Parts.

SECTION II.2. EXISTENCE THEOREMS OF ASYMPTOTIC SOLUTIONS TO SINGLE SYSTEMS.

In this section, we keep the notation used in Section I.2.

Let A_0 be an m-by-m constant matrix, and let $a(x,u)$ be an m-dimensional vectorial function holomorphic and strongly asymptotically developable as (x,u) tends to $H'=\{(x,u): x_1\cdots x_n=0\}$ in $S(c,r)\times D(R)=S(c_1,r_1)\times\ldots\times S(c_n,r_n)\times D(R_1)\times\ldots\times D(R_n)$. From the holomorphy with respect to u, $a(x,u)$ has the expansion of the form

$$(2.1)\qquad a(x,u) = a_0(x) + A(x)u + \sum_{q\in\mathbb{N}^m,\,|q|>1} a_q(x)u^q ,$$

where $a_0(x)$ and $a_q(x)$ $(q\in\mathbb{N}^m, |q|>1)$ are column vectors of functions holomorphic and strongly asymptotically developable in $S(c,r)$, and $A(x)$ is an m-by-m matricial function strongly asymptotically developable in $S(c,r)$. We suppose that

$$(2.2)\qquad \begin{cases} FA(a_0)(0) = TA(a_0)_{0_{[1,n]}} = \lim_{x\to 0} a_0(x) = 0 \\ FA(A)(0) = TA(A)_{0_{[1,n]}} = \lim_{x\to 0} A(x) = 0 . \end{cases}$$

Note that

$$(2.3)\qquad |(\partial/\partial u)a(x,u)| \leq |A(x)| + B|u|$$

for $(x,u)\in S[c',r']\times D[R']$, where $S[c',r']$ is a closed subpolysector of $S(c,r)$, $D[R']$ is a closed subpolydisc of $D(R)$, B is a constant dependent on the choice of $S[c',r']\times D[R']$, and $(\partial/\partial u)a(x,u)$ is the transposed matrix of $((\partial/\partial u_k)a(x,u))_{k=1,\ldots,m}$.

We shall consider following systems of equations

$$(E.1)\qquad A_0u = a(x,u) ,$$

$$(E.2)\qquad (yd/dy)u + A_0u = a(x,u) ,$$

$$(E.3)\qquad x^p(yd/dy)u + A_0u = a(x,u) ,$$

where we use y instead of x_1, and $p=(p_1,\ldots,p_n)\in \mathbb{N}^n$ with $p_1>0$. Our main objects to study are (E.2) and (E.3); we need (E.1) in order to construct formal solutions to (E.2) and (E.3).

If u is a solution of (E.1) (resp. (E.2) or (E.3)), holomorphic and strongly asymptotically developable in a subpolysector $S(c',r')$ of $S(c,r)$, then for any non-empty subset J of $[1,n]$, the formal series $u_J=FA_J(u)$ satisfies the following formal system $(E.1)_J$ (resp. $(E.2)_J$ or $(E.3)_J$)

$$(E.1)_J \qquad A_0u_J = b_J(x,u_J) ,$$

$$(E.2)_J \qquad (yd/dy)u_J + A_0u_J = b_J(x,u_J) ,$$

$$(E.3)_J \qquad x_J^{p_J}x_I^{p_I}(yd/dy)u_J + A_0u_J = b_J(x,u_J) ,$$

where $I=J^c$, $p=(p_J,p_I)$ and

$$(2.4)_J \qquad b_J(x,u_J) = FA_J(a_0) + FA_J(A)u_J + \sum_{q\in\mathbb{N}^m} FA_J(a_q)(u_J)^q .$$

Therefore, we have equations of $u(q_J)=TA(u)_{q_J}$ $(J\neq\emptyset, \subset[1,n], q_J\in\mathbb{N}^J)$ of the form $(E.1.q_J)$ (resp. $(E.2.q_J)$ or $(E.3.q_J)$)

$$(E.1.q_J) \qquad A_0u(q_J) = (q_J!)^{-1}(\partial/\partial x_J)^{q_J}(b_J(x,u_J(x)))\big|_{x_J=0} ,$$

$$(E.2.q_J) \qquad LHSR(q_J) = (q_J!)^{-1}(\partial/\partial x_J)^{q_J}(b_J(x,u_J(x)))\big|_{x_J=0} ,$$

where $LHSR(q_J)$ is defined by

$$LHSR(q_J) = (A_0+q_1I_m)u(q_J) \quad \text{if} \quad 1\in J ,$$

$$LHSR(q_J) = (yd/dy)u(q_J) + A_0u(q_J) \quad \text{if} \quad 1\notin J ,$$

$$(E.3.q_J) \qquad LHS(q_J) = (q_J!)^{-1}(\partial/\partial x_J)^{q_J}(b_J(x,u_J(x)))\big|_{x_J=0} ,$$

where, left-hand side is defined according to J as follows: denoting by $J(+p)$ the set $\{j\in[1,n];\ p_j>0\}$, put

$$LHS(q_J) = A_0u(q_J) \quad \text{if} \quad J(+p)\cap J\neq\emptyset .$$

$$LHS(q_J) = x_I^{p_I}(yd/dy)u(q_J) + A_0 u(q_J) \quad \text{if} \quad J(+p) \cap J = \emptyset .$$

By calculation, we can verify that the right-hand side is equal to

$$\begin{cases} TA(a_0)_{0_J} + TA(A)_{0_J} u(0_J) + \sum_{q \in \mathbb{N}^m, |q|>1} TA(a_q)_{0_J} u(0_J)^q , & \text{if } |q_J|=0 \text{ i.e. } q_J=0_J, \\ ((\partial/\partial u)(a(x,u))\big|_{u=u(0_J), x_J=0})u(q_J) + \text{terms determined by} & \end{cases}$$

$\{a_0, A, a_q\ (|q|>1)$ and $u(b_J)(|b_J|<|q_J|)\}$, if $|q_J|>0$.

DEFINITION 2.1. Let

$$F = \{f(x_I;q_J);\ J \subset [1,n],\ J \neq \emptyset,\ I=J^c,\ q_J \in \mathbb{N}^J\}$$

be a consistent family in $S(c',r')$ (resp. strictly consistent family in $S[c',r']$). We say that the family F is a family of formal solutions of (E.1) (resp. (E.2) or (E.3)), if for any non-empty set J of $[1,n]$, the power series

$$PS(F) = \sum_{q_J \in \mathbb{N}^J} f(x_I;q_J) x_J^{q_J}$$

satisfies in $S_I = \Pi_{i \in I} S(c_i',r_i')$ (resp. $=\Pi_{i \in I} S[c_i',r_i']$) the system obtained from $(E.1)_J$ (resp. $(E.2)_J$ or $(E.3)_J$) by replacing u_J with $PS(F)_J$, namely, $f(x_I;q_J)$ satisfies the equation obtained from $(E.1.q_J)$ (resp. $(E.2.q_J)$ or $(E.3.q_J)$) by replacing $u(q_J)$ with $f(x_I;q_J)$ for any $J \neq \emptyset$ and any $q_J \in \mathbb{N}^J$.

In the case where $f(x_I;q_J)$ is holomorphic in the disc $D_I(r') = \Pi_{i \in I} D(r_i')$, we say that $PS(F) = \sum_{q \in \mathbb{N}^n} f(x_\emptyset;q)x^q$ is a formal power-series solution in $\mathcal{O}_{D(r')} \hat{|}_H (D(r'))$.

Let F be a strictly consistent family of formal solutions of (E.1) (resp. (E.2) or (E.3)) in $S(c',r')$. For any $N \in \mathbb{N}^n$, we denote by $V_N[m,c',r']$ the set of all column vectors $u=(u_j)_{j \in [1,m]}$ of functions holomorphic in $S[c',r']$, and

$$|f|_{N,c',r'} = \max_{j \in [1,m]} \sup\{|x|^{-N}|f_j(x)|: x \in S[c',r']\}$$

is finite. Then, a column vector $u=(u_j)_{j\in[1,m]}$ in $V_N[m,c',r']$ is a solution of the system (E.1) (resp. (E.2) or (E.3)) such that u is strictly strongly asymptotically developable in $S[c',r']$ and the total family TA(u) of coefficients of strictly strongly asymptotic expansion coincides with the given family F of formal solutions, if, for any $N\in\mathbb{N}^n$, $u-App_N(F)$ belongs to $V_N[m,c',r']$, where $App_N(F)$ is defined by (I.2.9) with the family F.

Consider now the system (E.1).

PROPOSITION 2.1. *Let* $F=\{u(q_J)\}$ *be a strictly consistent family of formal solutions of* (E.1) *with* $u(0_{[1,n]})=0$ *in a closed subpolysector* $S[c',r']$. *If* A_0 *is invertible, then* (E.1) *has a unique solution* u *in* $S[c',r'']$ *which is strictly strongly asymptotically developable in* $S[c',r'']$ *with* TA(u)=F.

PROOF. Choose r" so small that

$$\sup\{|u(0_J)|;\ x_I\in\Pi_{i\in I}S[c_i',r_i''],\ I=J^c\}\leq 2^{-n-2}R'$$

for any non-empty subset J of [1,n], and

$$\sup\{|A_0^{-1}A|;\ x\in S(c',r'')\}\leq 4^{-1}\ .$$

Let $v\in V_0[m,c',r'']$, and put $Lv=A_0^{-1}a(x,v(x))$. We shall prove that for any $N\in\mathbb{N}^n$, there exists a constant K_N such that the operator L_N defined by

$$L_Nv = L(v+App_N(F)) - App_N(F)$$

is a contraction operator of $V_{N,K_N}[m,c',r_N]$, where r_N is chosen so that $r_{N,i}=r_i''$ if $N_i=0$. We see that L_N is written in the form

$$L_Nv = A_0^{-1}a(x,g_N) - g_N + A_0^{-1}(a(x,v+g_N)-a(x,g_N))$$

where $g_N=App_N(F)$. As F is a strictly consistent family of formal solutions, we can estimate

$$|A_0^{-1}a(x,u+g_N)-g_N|\leq C_N|x|^N\ ,$$

for any $x \in S[c',r'']$. Choose r_N and K_N so that

$$\begin{cases} K_N r_N{}^N \leq 2^{-1}R' , \\ |g_N - \sum_J (-1)^{\#J+1} u(0_J)| \leq 4^{-1}R' , \\ (|A_0{}^{-1}|BK_N r_N{}^N + 4^{-1})K_N + C_N \leq K_N , \\ 2|A_0{}^{-1}|BK_N r_N{}^N + 4^{-1} \leq 2^{-1} . \end{cases}$$

In fact, we can put $K_N = 8C_N/5$ and choose $r_{N,i}$ $(N_i \neq 0)$ enough small to obtain the first, second and fourth inequalities. Then, using (2.3), we see that L_N is a contraction operator of $V_{N,K_N}[m,c',r_N]$. Then, by the principle of contraction operator, there exists a column vector u_N in $V_{N,K_N}[m,c',r_N]$ satisfying $L_N u_N = u_N$, and such a column vector is unique for any $N \in \mathbb{N}^n$. The definition of L_N means that $f_N \overset{\text{def}}{=\!=} u_N + App_N(F)$ satisfies $Lf_N = f_N$, for any $N \in \mathbb{N}^n$. Moreover, for any N, $N' \in \mathbb{N}^n$, $N > N'$, $f_N - App_{N'}(F)$ belongs to $V_{N',K_{N'}}[m,c',\min\{r_N,r_{N'}\}]$ and $f_N - App_{N'}(F)$ is a solution to $L_{N'}v = v$. By the uniqueness of solution, $f_N - App_{N'}(F) = u_{N'}$ i.e. $f_N = f_{N'}$. Define a vectorial function u in $S[c',r'']$ (note that $r' = \sup\{r_N : N \in \mathbb{N}^n\}$) by

$$u(x) = f_N(x) \quad (x \in S[c',r_N],\ N \in \mathbb{N}^n) .$$

Then, by Lemma I.2.1, u is strictly strongly asymptotically developable in $S[c',r'']$ with $TA(u) = F$. Q.E.D.

REMARK 2.1. If $a(x,u) = a_0(x) + A(x)u$, then we require only the inequalities

$$|A_0{}^{-1}A| \leq 4^{-1} \quad \text{and} \quad C_N + 4^{-1}K_N \leq K_N .$$

PROPOSITION 2.2. *If A_0 is invertible, then, for any closed subset $c' = \prod_{i=1}^n c_i'$ of c there exists a unique strictly consistent family $F = \{u(q_J)\}$ of formal solutions to (E.1) in $S[c',r']$ with $u(0_{[1,n]}) = 0$, where r' is a constant adequately chosen.*

For this family F, the statement of PROPOSITION 2.1 is valid.

PROOF. Suppose that $F = \{u(q_J)\}$ is a strictly consistent family of formal solutions of (E.1). Then, we obtain the equations $(E.1.q_J)$'s $(J\neq\emptyset,\ J\subset[1,n],\ q_J\in\mathbb{N}^J)$. We shall solve these equations in an inductive way on the cardinal number $\#J$ of J and the length $|q_J|$ of index q_J. In the case $\#J=n$, $(E.1.q_J)$'s are inductively solved because A_0 is invertible. Suppose that we have a unique solution $u(q_J)$ of $(E.1.q_J)$ for any J, $\#J>n'$, $q_J\in\mathbb{N}^J$, such that

$$\{u(q_{J'\cup J});\ J'\subset I=J^c,\ J'\neq\emptyset\}$$

is a strictly consistent family of formal solutions $(E.1.q_J)$ for $\#J=n'$ and $q_J\in\mathbb{N}^J$ in $\Pi_{i\in I}S[c_i',r_i(q_J)]$. Then, we apply PROPOSITION 2.1 to these family of formal solutions of $(E.1.q_J)$ inductively on $|q_J|$. Because the equations $(E.1.q_J)$'s $(|q_J|>0)$ are linear, we have a constant $r_{n'}$ such that the solution $u(q_J)$ is defined in $\Pi_{i\in I}S[c_i',r_{n',i}]$ for any J, $\#J=n'$ and $q_J\in\mathbb{N}^J$. By the construction, we can verify evidently the consistency. Q.E.D.

<u>Corollary 2.1</u>. <u>If</u> A_0 <u>is invertible and if</u> $a(x,u)$ <u>is strongly asymptotically developable to an element in</u> $\mathcal{O}_{D(r)\times D(R)}\hat{|}H'(D(r)\times D(R))^m$, <u>then there exists a unique formal power-series solution</u> $\hat{u}$ <u>in</u> $\mathcal{O}_{D[r']}\hat{|}H(D[r'])^m$ <u>satisfying</u> $\hat{u}(0)=0$ <u>and a unique solution</u> u <u>holomorphic and strongly asymptotically developable to</u> $\hat{u}$ <u>in</u> $S[c',r'']$, <u>where</u> $r''\leq r'<r$ and $c'\subset c$.

<u>Corollary 2.2</u>. <u>If</u> A_0 <u>is invertible and if</u> $a(x,u)$ <u>is holomorphic in</u> $D(r)\times D(R)$, <u>then there exists a unique formal power-series solution</u> u <u>in</u> $\mathcal{O}_{D[r']}\hat{|}H(D[r'])$ <u>satisfying</u> $u(0)=0$. <u>Morevoer, the formal power-series solution</u> u <u>is convergent in</u> $D[r']$, where r' is sufficiently small.

Secondly, consider the system (E.2).

<u>THEOREM.2.1</u>. <u>Suppose that</u> $F=\{u(q_J)\}$ <u>is a strictly consistent family of formal solutions to</u> (E.2) <u>with</u> $u(0_{[1,n]})$ <u>in a closed subpolysector</u> $S[c',r']$ <u>of</u> $S(c,r)$. <u>Then,</u> (E.2) <u>has a unique</u> u <u>in</u> $V_0[m,c',r'']$ <u>which is strictly strongly asymptotically developable in</u> $S[c',r'']$ <u>and whose total family of coefficients of asymptotic expansion coincides with the family</u> F <u>of formal solutions, where</u> r'' <u>is some constant less than or equal to</u> r'.

PROOF. Choose r" so that

$$|u(0_J)| \leq 2^{-n-2}R' ,$$

for any $x \in S(c',r")$ and any non-empty subset J of [1,n], and

$$|A_0| + |A| \leq 4^{-1} .$$

For $u \in V_0(m,c,r)$, set

$$Lu = \int_0^1 \{a(ty,x',u(ty,x')) - A_0 u(ty,x')\} t^{-1} dt$$

and set

$$L_N u = L(u+App_N(F)) - App_N(F) .$$

The operator L_N is written in the form

$$L_N u = \int_0^1 E_N(ty,x') t^{-1} dt + \int_0^1 F_N(ty,x',u(ty,x')) t^{-1} dt$$

where

$$E_N(x) = a(x,App_N(F)) - A_0 App_N(F) - (y\partial/\partial y) App_N(F) ,$$

$$F_N(x,u) = a(x,u+App_N(F)) - a(x,App_N(F)) - A_0 u .$$

By the assumption, we have a constant C_N such that

$$|E_N(x)| \leq C_N |x|^N$$

for any $x \in S[c',r"]$. Put $N_0=(1,0,\ldots,0)$. For any $N \geq N_0$, we choose K_N and r_N so that

$$\begin{cases} K_N r_N^{\ N} \leq 2^{-1}R' , \\ |App_N(F) - \sum_J (-1)^{\#J+1} u(q_J)| \leq 4^{-1}R' , \end{cases}$$

$$\begin{cases} C_N + (4^{-1}+BK_N r_N{}^N)K_N \leq K_N N_1 \ , \\ 4^{-1} + 2BK_N r_N{}^N \leq 2^{-1}N_1 \ , \\ r_{N,i} = r_i'' \quad \text{if} \quad i>1 \text{ and } N_i=0, \text{ or } i=1 \text{ and } N_1=1 \ . \end{cases}$$

Then, L_N is a contraction of $V_{N,K_N}[m,c',r_N]$ for any $N>N_0$.

By the principle of contraction operator, there exists a column vector u_N in $V_{N,K_N}[m,c',r_N]$ satisfying $L_N u_N=u_N$, and such a column vector is unique for any $N \geq N_0$. Put $f_N=u_N+App_N(F)$. Then, $Lf_N=f_N$. Moreover, for any N, $N' \geq N_0$, if $N>N'$, $f_N-App_N(F)$ belongs to $V_{N',K_{N'}}[m,c',\min\{r_N,r_{N'}\}]$ and $f_N-App_{N'}(F)$ is a solution to $L_{N'}v=v$. Therefore, by the uniqueness of solution, $f_N-App_{N'}(F)=u_{N'}$, i.e. $f_N=f_{N'}$. Hence, we can define a vectorial function u in $S[c',r'']$ (note that $r''=\sup\{r_N:N \geq N_0\}$) by

$$u(x) = f_N(x) \qquad (x\in S[c',r_N],\ N \geq N_0) \ .$$

By Lemma I.2.1, u is strictly strongly asymptotically developable in $S[c',r'']$ with $TA(u)=F$. Q.E.D.

REMARK 2.2. In the case $a(x,u)=a_0(x)+A(x)u$, to prove THEOREM 2.1, we can choose r'' and K_N so that

$$|A_0 + A| \leq 4^{-1} \quad \text{and} \quad C_N + 4^{-1}K_N \leq N_1 K_N \ .$$

THEOREM 2.2. *If A_0+KI_m is invertible for any nonnegative integer k, then there exists only one family F of formal solutions to (E.2) satisfying $PS(F)(0)=0$ in any closed subpolysector $S[c',r']$ of $S(c,r)$ with r' adequately chosen.*

For this family F, the statement in THEOREM 2.1 is valid.

PROOF. If $F=\{u(q_J)\}$ is a family of formal solutions of (E.2), then for any non-empty subset J of $[1,n]$ and any $q_J\in \mathbb{N}^J$, $u(q_J)$ satisfies the equation (E.2.q_J). We shall prove that these equations can be solved in an inductive way on $\#J$ and $|q_J|$. In the case $\#J=n$, (E.2.q_J)'s can be solved inductively because $A_0+q_1I_m$ ($q_1\in \mathbb{N}$) are invertible. Suppose that we have solutions

$$\{u(q_J);\ \#J > n',\ q_J \in \mathbb{N}^J\}\ ,$$

in $S(c', r_{n'+1})$. To the family of formal solutions of $(E.2.q_J)$

$$\{u(q_{J' \cup J});\ J' \subset I = J^c,\ J' \neq \emptyset\}\ ,$$

for $\#J = n'$, $q_J \in \mathbb{N}^J$, we can apply PROPOSITION 2.1 if $1 \in J$, and otherwise THEOREM 2.1. Hence, we obtain solutions $u(q_J)$'s in $\Pi_{i \in I} S[c_i', r_i'']$'s respectively. By the construction, $F = \{u(q_J)\}$ is a consistent and an unique family of formal solutions of (E.2). Q.E.D.

Corollary 2.3. *suppose that* $a(x,u)$ *is strongly asymptotically developable to an element in* $\mathcal{O}_{D(r)\times D(R)}\hat{|}_{H'}(D(r)\times D(R))^m$. *If* (E.2) *has a formal power-series solution* $\hat{u}$ *in* $\mathcal{O}_{D[r']}\hat{|}_H(D[r'])^m$ *satisfying* $\hat{u}(0)=0$, *then there exists a unique solution* u *holomorphic and strongly asymptotically developable to* $\hat{u}$ *in* $S[c', r'']$, *where* $r'' \leq r' < r$ *and* $c' \subset c$. *Moreover, if* $A_0 + kI_m$ *is invertible for any nonnegative integer, then there exists a unique solution* u *holomorphic and strongly asymptotically developable to an element in* $\mathcal{O}_{D[r']}\hat{|}_H(D[r'])^m$ *in* $S[c', r'']$ *satisfying* $\lim_{x\to 0} u(x) = 0$.

Corollary 2.4. *Suppose that* $a(x,u)$ *is holomorphic in the disc* $D(r)\times D(R)$. *If* (E.2) *has a formal power-series solution* $\hat{u}$ *in* $\mathcal{O}_{D[r'}\hat{|}_{H]}(D[r'])$, *then* $\hat{u}$ *is convergent in* $D[r']$, *where* $r' < r$. *Moreover, if* $A_0 + kI_m$ *is invertible for any nonnegative integer, then there exists a unique convergent power-series solution* u *to* (E.2) *with* $u(0)=0$.

Consider finally the system (E.3). If $a(x,u)=0$, then $W(x)=\exp(p_1^{-1}A_0x^{-p})$ is a matrix of solution and a solution $u(x)$ of (E.3) satisfies the integral equation

$$u(x) = W(x)\{W(y_0,x')^{-1}u(y_0,x') + \int_{y_0}^{y} W(x)^{-1}a(x,u(x))x^{-p}y^{-1}dy\}\ .$$

We can suppose, without loss of generality, that $A_0 = D + T$ with a diagonal matrix $D = \oplus_{j=1}^{m} d_j$ and a properly upper triangular matrix T commutative with D. Then, the integral equation is written in the form

$$u(x) = U(x)\{U(y_0,x')^{-1}\ u(y_0,x') + \int_{y_0}^{y} u(s,x')b(s,x')x'^{-p'}s^{-p_1-1}ds\}$$

where $U(x)=\oplus_{j=1}^{m}U_j(x)$, $U_j(x)=\exp(p_1{}^{-1}d_j x^{-p})$, $b(x)=a(x,u(x))-Tu(x)$ and $p'=(p_2,\ldots,p_n)$. In order to prove that the solution $u(x)$ is strongly asymptotically developable as x tends to H in $S(c,r)$, we shall choose y_0, $u(y_0,x')$ and path of integration adequately according to the asymptotic behavior of $|U_j(x)|$'s. Thus, we come to define proper domains of $|U(x)|$ as follows.

DEFINITION 2.2. Let μ be a non-zero complex number. A subset set c' of $\mathbb{R}^n$ or T^n is proper with respect to the function $|\exp(\mu x^{-p})|=\exp(|\mu|\cos(\arg\mu-(p,\arg x))|x|^{-p})$, if the set

$$c'\cap\{o\in\mathbb{R}^n(\text{or } T^n);\ \cos(\arg\mu-(p,o))>0\}$$

has at most one connected component, where $(p,o)=\sum_{i=1}^{n}p_i o_i$.

DEFINITION 2.3. A connected subset c' of $\mathbb{R}^n$ (or T^n) is strictly proper with respect to $|\exp(\mu x^{-p})|$, if there exists a positive constant e such that the set

$$c'\cap\{o\in\mathbb{R}^n\ (\text{or } T^n);\ \cos(\arg\mu-(p,o))>-\sin 4e\}$$

has at most one connected component, namely, for any $o\in c'$

$$4e-3\pi/2<\arg\mu-(p,o)+2k\pi<-4e+3\pi/2$$

with some integer k. If c' is sufficiently small, then c' is strictly proper with respect to the function.

DEFINITION 2.4. Suppose that the matrix D is invertible. A connected subset c' of $\mathbb{R}^n$ (or T^n) is said to be proper (resp. strictly proper) with respect to $|U(x)|$ if c' is proper (resp. strictly proper) with respect to $|U_j(x)|$ for all $j=1,\ldots,m$.

Let c' be a strictly proper domain with respect to $|\exp(\mu x^{-p})|$, and put

$$w_-=\inf\{(p,o);\ o\in c'\}\ ,\qquad w_+=\sup\{(p,o);\ o\in c'\}\ ,$$

$$w_r=\arg\mu+\pi/2\quad\text{and}\quad w_l=\arg\mu-\pi/2\ .$$

Then, there exists a positive constant e such that one of the following inequalities is valid modulo $\mathbb{Z}$ for all $o \in c'$:

$$(c-) \qquad w_1-\pi+4e \leq w_- \leq (p,o) \leq w_r-4e \ ,$$

$$(c+) \qquad w_1+2e \leq w_- \leq (p,o) \leq w_+ \leq w_r-2e \ ,$$

$$(c-+) \qquad w_1-\pi+4e \leq w_- \leq w_1-2e \leq (p,o) \leq w_r-2e \leq w_+ \leq w_r \ ,$$

$$(c+-) \qquad w_1 \leq w_- \leq w_1+2e \leq (p,o) \leq w_r+2e \leq w_+ \leq w_r-4e \ ,$$

or

$$(c-+-) \qquad w_1-\pi+4e \leq w_- \leq w_1-2e \leq (p,o) \leq w_r+2e \leq w_+ \leq w_r+\pi-4e \ .$$

DEFINITION 2.5. We say that c' is a negatively strictly proper briefly negative domain with respect to $|\exp(\mu x^{-p})|$ if all elements in c' satisfy the inequality (c−). In the other case, we say that c' is non-negatively strictly proper, briefly non-negative domain with respect to $|\exp(\mu x^{-p})|$.

Notice that $\exp(\mu x^{-p})$ is strictly strongly asymptotically developable to 0 in S[c',r'] if c' is negatively strictly proper domain with respect to $|\exp(\mu x^{-p})|$.

If c' is a strictly proper domain with respect to $|U(x)|$, then we denote by NI(c,U) the subset of all j such that c' is a negatively strictly proper domain with respect to $|U_j(x)|$.

For a strictly proper domain c' with respect to $|\exp(\mu x^{-p})|$, we define functions h_+ and h_- of $s=(p,o)$ $(o \in c')$ as follows: if c' is negative, choose s_0 in (w_-,w_+) and

$$h_-(s) = s-w_1+2e \qquad (w_- \leq s < s_0)$$

$$h_-(s) = 2e \qquad (s = s_0)$$

$$h_-(s) = s-w_r+2e \qquad (s_0 < s \leq w_+)$$

$$h_+(s) = s-w_1-2e+\pi \qquad (w_- \leq s < s_0)$$

$$h_+(s) = \pi-2e \qquad (s=s_0)$$

$$h_+(s) = s-w_r-2e+\pi \qquad (s_0 < s \leq w_+) ,$$

if c' is nonnegative,

$$h_-(s) = s-w_1+2e \qquad (w_- \leq s < \max\{w_-, w_1+4e\})$$

$$h_-(s) = 2e \qquad (\max\{w_-, w_1+4e\} \leq s \leq \min\{w_+, w_r-4e\})$$

$$h_-(s) = s-w_r+2e \qquad (\min\{w_+, w_r-4e\} < s \leq w_+)$$

$$h_+(s) = s-w_1-2e+\pi \qquad (w_- \leq s < \max\{w_-, w_1+4e\})$$

$$h_+(s) = \pi-2e \qquad (\max\{w_-, w_1+4e\} \leq s \leq \min\{w_+, w_r-4e\})$$

$$h_+(s) = s-w_r-2e+\pi \qquad (\min\{w_+, w_r-4e\} < s \leq w_+) .$$

If c' is a strictly proper domain with respect to $|U(x)|$, we define $w_{1,j}$, $w_{r,j}$, $h_{-,j}$ and $h_{+,j}$ as avove for all $j=1,\ldots,m$, and put

$$h_-(s) = \max\{h_{-,j}(s);\ j=1,\ldots,m\} ,$$

$$h_+(s) = \min\{h_{+,j}(s);\ j=1,\ldots,m\} .$$

LEMMA 2.1. *For any* $s=(p,o)$ $(o \in c')$, *the following inequalities are valid:*

$$h_-(s) < \pi,\ h_+(s) > 0 \text{ and } h_-(s) \leq h_+(s) .$$

PROOF. The first two inequalities are evidently satisfied by the definitions. We shall prove $h_{-,j'}(s) \leq h_{+,j}(s)$ for any two numbers $i,j' \in [1,m]$. If the intervals $(w_{j,1}, w_{j,r})$ and $(w_{j',1}, w_{j',r})$ are included in the interval (w_-, w_+), then

$$w_{1,j}-w_{1,j'} < \pi-4e \quad\text{and}\quad w_{r,j}-w_{r,j'} < \pi-4e .$$

This implies the inequality $h_{-,j'}(s) \leq h_{+,j}(s)$. In a similar way, we can verify this inequality for the other cases. Q.E.D.

From the above lemma, we can find a function $t((p,o))$ on c' such that

$$0 < t((p,o)) < \pi, \text{ and } h_-((p,o)) \leq t((p,o)) \leq h_+((p,o)) .$$

Using this function, we choose paths of integration as follows: if c' is a negatively strictly proper domain with respect to $|U_j(x)| = |\exp(d_j p_1^{-1} x^{-p})|$, then choose x_0 satisfying $s_0 = (p, \arg x_0)$ and define a path $r_j(x) = (y_j(\sigma), x')_{s \in [0,2]}$ jointing $(x_{0,1}, x')$ to x by

$$y_j(s) = (1-s)x_{0,1} + s\{y \exp(\int_2^1 \cot(t(f(\sigma))df(\sigma) + (-1)^{1/2} f(1))\}$$

for $s \in [0,1]$, and

$$y_j(s) = y\{\exp(\int_2^s \cot(t(f(\sigma))df(\sigma) + (-1)^{1/2} f(1))\}$$

for $s \in [1,2]$, where $f(s) = p_1((s-1)\arg y + (2-s)\arg x_{0,1}) + (p', \arg x')$, and if c' is a non-negatively proper domain with respect to $|U_j(x)|$, then define a path $r_j(x) = (y_j(s), x')_{s \in [0,2]}$ from $(0,x')$ to x such that

$$w_- \leq (p, \arg x) < \max\{w_-, w_1 + 4e\} \text{ or } \min\{w_+, w_r - 4e\} < (p, \arg x) \leq w_+ ,$$

by the piecewise continuous curve

$$y_j(s) = sy\{\exp(\int_2^1 \cot(t(f(\sigma))df(\sigma) + (-1)^{1/2} f(1))\} \quad (s \in [0,1]) ,$$

$$y_j(s) = y\{\exp(\int_2^s \cot(t(f(\sigma))df(\sigma) + (-1)^{1/2} f(1))\} \quad (s \in [1,2]) ,$$

where $f(1) = w_1 + 4e$ or $w_r - 4e$, and

$$f(s) = p_1((s-1)\arg x_{0,1} + (2-s)f(1)) + (p', \arg x') ,$$

and for x such that

$$\max\{w_-, w_1 + 4e\} \leq (p, \arg x) \leq \min\{w_+, w_r - 4e\} ,$$

by the straight line

$$y_j(s) = (sy)/2 .$$

LEMMA 2.2. *Suppose that D is invertible and keep the above notation. For any* $N=(N_1,N')\in\mathbb{N}^n$ *and for any* $r=(r_1,r')\in(\mathbb{R}^+)^n$ *satisfying*

$$(2.5)\quad N_1(r_1^{mx})^{p_1}r'^{p'}\leq |d_j|(\sin 2e-\sin e)\quad (j=1,\dots,m),$$

we obtain the estimates

$$|U_j(x)|\left|\int_{r_j(x)}|U_j(y_j(s),x')|^{-1}|x'|^{N'-p'}|Y_j(s)|^{N_1-1-p_1}|dy_j(s)|\right.$$

$$\leq (|d_j|\sin e)^{-1}|x|^N\quad (j=1,\dots,m),$$

for any x *in the sectorial domain* $S[c',r',t]$ *defined by*

$$S[c',r',t]$$

$$=\{x\in\mathbb{C}^n;\arg x\in c',\ 0<|x'|\leq r',\ 0<|y|\leq r_1\exp(\int_{w_-}^{(p,\arg x)}\cot(t(s))ds)\},$$

where $x_{0,1}=r_1\exp(\int_{w_-}^{(p,\arg x_0)}\cot(t(s))ds)$ *and*

$$r_1^{mx}=r_1\sup\{\exp\int_{w_-}^{(p,o)}\cot(t(s))ds;\ o\in c'\}.$$

PROOF. We shall prove the estimates

$$(d/ds)\{|y_j(t)|^{N_1}|x'|^{N'}|U_j(s)|^{-1}\}$$

$$\geq |d_j|\sin e|U_j(y_j(s),x')||x'|^{N'-p'}|y_j(s)|^{N_1-p_1-1}|dy_j(s)/ds|,$$

on the paths $r_j(x)$, $j=1,\dots,m$. By calculating the differential,

$$|U_j(s)|(d/ds)\{|y_j(t)|^{N_1}|x'|^{N'}|U_j(s)|^{-1}$$

$$=\{N_1|y_j(s)|^{N_1-1}|x'|^{N'}(d|y_j(s)|/ds)$$

$$-|d_j||y_j(s)|^{N_1-1-p_1}|x'|^{N'-p'}\sin(\arg(p_1^{-1}d_j)-h(s))(df(s)/ds)$$

$$+|d_j||y_j(s)|^{N_1-1-p_1}|x'|^{N'-p'}\cos(\arg(p_1^{-1}d_j)-h(s))(d|y_j(s)|/ds)$$

$$=\{N_1|y_j(s)|^{p_1}|x'|^{p'}(d|y_j(s)|/ds)$$

$$-|d_j||y_j(s)|^{-1}\sin(\arg(p_1^{-1}d_j)-h(s))(df(s)/ds)$$

$$+|d_j|\cos(\arg(p_1^{-1}d_j)-h(s))(d|y_j(s)|/ds)\}$$

$$\times\{|y_j(s)|^{N_1-1-p_1}|x'|^{N'-p}\} ,$$

where $h(s)=p_1\arg y_j(s)+(p',\arg x')$.

Suppose that c' is nonnegative with respect to $|U_j(x)|$ and that x satisfies

$$\mathrm{Max}\{w_-,w_1+4e\}\leq(p,\arg x)\leq\min\{w_+,w_r-4e\} ,$$

then the desired inequality is obviously valid, because the path is the line jointing $(0,x')$ to x.

Suppose that c' is nonnegative with respect to $|U_j(x)|$ and that x satisfies

$$w_-\leq(p,\arg x)<\max\{w_-,w_1+4e\} \text{ or } \min\{w_+,w_r-4e\}<(p,\arg x)\leq w_+ .$$

Then, by the same reason as above, for $s\in[0,1]$ the inequality is valid; for $s\in[1,2]$, we can verify that

$$d|y_j(s)|/ds = |y_j(s)|\cot(t(f(s))(df(s)/ds)$$

and

$$|dy_j(s)/ds| = |y_j(s)|(1+\cot^2 t(f(s)))^{1/2}|df(s)/ds|$$

$$= |y_j(s)|\sin t(f(s))|df(s)/ds| .$$

Therefore, the desired inequality is written in the form

$$\{|d_j|\cos(\arg(p_1^{-1}d_j)-h(s)+t(f(s)))+N_1|y_j(s)|^{p_1}|x'|^{p'}|df(s)/ds)|\}$$

$$\geq|d_j|(\sin e)|df(s)/ds| .$$

Because of (2.5), it suffices to prove the following inequality

$$\cos(\arg(p_1^{-1}d_j)-h(s)+t(f(s)))(df(s)/ds)\geq(\sin 2e)|df(s)/ds|$$

which is satisfied by the definition of $t(f(s))$.

Suppose finally that c' is negative with respect to $|U_j(x)|$. For $s \in [1,2]$, by the same argument as the second case, we see that the inequality is valid. For $s \in [0,1]$, by the definition, we see that

$$df(s)/ds = 0, \quad d|y_j(s)|/ds = -|dy_j(s)/ds| < 0 ,$$

and so, it suffices to prove the inequality

$$-|d_j|\cos(\arg(p_1{}^{-1}d_j)-h(s))-N_1|y_j(s)|^{p_1}|x'|^{p'} \geq |d_j|\sin e .$$

Because of (2.5), we have only to prove

$$-\cos(\arg(p_1{}^{-1}d_j)-h(s)) \geq \sin 2e .$$

As c' is negative with respect to $|U_j(x)|$, this inequality is valid. Q.E.D.

For any $N \in \mathbb{N}^n$, denote by $V_N[m,c',r_N,t]$ the set of all m dimensional column vectors f of holomorphic functions in the sectorial domain $S[c',r_N,t]$ such that the norm of f

$$|f|_{N,c',r_N} = \sup\{|f(x)||x|^{-N};\ x \in S(c',r_N,t)\}$$

is finite.

<u>THEOREM 2.3</u>. <u>We assume that</u> $A_0=D+T$ <u>is invertible</u>. <u>Let</u> $c'=\Pi_{i=1}^{n}c_i{}'$ <u>be a strictly proper domain in c with respect to</u> $|U(x)|$, <u>and suppose that</u> $F=\{u(q_J)\}$ <u>is a strictly consistent family of formal solutions to</u> (E.3) <u>in</u> $S[c',r']$ <u>with</u> $u(0_{[1,n]})=0$. <u>Let</u> $v_j(x')$'s $(j \in NI(c',U))$ <u>be functions such that, for any subset</u> J' <u>of</u> $[2,n]$ <u>and</u> $q_{J'} \in \mathbb{N}^{J'}$,

$$TA(v_j)_{q_{J'}} = (u(q_{J'})(x_{0,1},x_{I'}))_j \quad (j \in NI(c',U)) ,$$

<u>where</u> $I'=J'^c$ <u>in</u> $[2,n]$ <u>and</u> $x_{0,1}$ <u>is sufficiently small</u>.

<u>Then, there exists a unique solution</u> $u(x)$ <u>of (E.3) in</u> $V_0[m,c',r'',t]$, <u>such that</u> u <u>is strictly strongly asymptotically developable in</u> $S[c',r'',t]$ <u>with TA(u)=F, and that for</u> $j \in NI(c',U)$

$$u_j(x_{0,1},x') = v_j(x') ,$$

<u>where</u> $0 < r'' \leq r'$.

PROOF. For the strictly proper domain c', we define the function $t((p,o))$ as above. Put $d=\min\{|d_j|;\ j \in [1,m]\}$.

If necessary, by using a linear transformation

$$u = \mathrm{diag}(1,q,\ldots,q^{n-1})v,$$

with q enough small, we can suppose that $|T| \leq 8^{-1}(d \sin e)$.

At first, choose $r'', x_{0,1}, K_0$ and C_0 such that

$r'' \leq r'$ (hence, (2.3) holds) ,

$u(0_J) \leq 2^{-n-2}R'$, for any J and $x_I \in \Pi_{i \in I} S[c_i',r_i'']$,

$(d \sin e)^{-1}|A(x)| \leq 4^{-1}$, for any $x \in S[c',r'']$,

$|a(x,0)| \leq C_0$, for any $x \in S[c',r'']$,

$|U_j(x)U_j(x_{0,1},x')^{-1}v_j(x')| \leq C_0$, for $j \in NI(c',U)$ and for any $x \in S[c',r'']$,

$K_0 \leq 2^{-1}R'$,

$C_0(1+(d \sin e)^{-1})+(4^{-1}+(\sin e)^{-1}BK_0)K_0 \leq K_0$,

and $(4^{-1}+2B(d \sin e)^{-1}K_0) \leq 2^{-1}$.

This choice is possible, because of (2.2) and $u(0_{[1,n]})=0$ and because $U_j(x)$ is strictly strongly asymptotic to 0 in $S[c',r'']$. For $f \in V_0[m,c',r'',t]$, set

$$(*)\quad (Lf)_j = U_j(x)\int_{r_{j,0}(x)} U_j(y_j(s),x')^{-1}x'^{-p'}y_j(s)^{-p_1-1}b_j^0(y_j(s),x':f)dy_j(s)$$

for $j \notin NI(c',U)$ and

(**) $(Lf)_j = U_j(x)U_j(x_{0,1},x')^{-1}v_j(x')$

$$+U_j(x)\int_{r_{j,0}(x)} U_j(y_j(s),x')^{-1}x'^{-p'}y_j(s)^{-p_1-1}b_j^0(y_j(s),x':f)dy_j(s)$$

for $j\in NI(c',U)$, where $(b_j^0(x:u))_{j\in[1,m]}=a(x,u)-Tu$ and $r_{j,0}(x)$ is defined as above with $t((p,o))$ and $x_{0,1}$. Then, by the choice of constants, L is a contraction operator of $V_{0,K_0}[m,c',r'',t]$. Therefore, there exists a unique $u_0\in V_{0,K_0}[m,c',r'',t]$ satisfying $Lu_0=u_0$.

For any $N\in\mathbb{N}^n$, define $E_N(x)$ and $F_N(x,u)$ by

$$E_N(x) = a(x,App_N(F))-A_0App_N(F)-x^p(yd/dy)App_N(F)$$

$$F_N(x,u) = a(x,u+App_N(F))-a(x,App_N(F)) .$$

By the definition, E_N is strictly strongly asymptotically developable and the approximate function $App_N(E_N)=0$, and so we have an estimate

$$|E_N(x)|\le C_N|x|^N$$

in the sector $S[c',r'']$. As $U_j(x)$ $(j\in NI(c',U))$ is strictly strongly asymptotic to 0 in $S[c',r'']$, we have an estimate

$$|U_j(x)U_j(x_{N,1},x')(u_{0,j}(x_{N,1},x')-App_N(F)(x_{N,1},x')_j)|\le C_N|x|^N ,$$

in the sector $S[c',r'']$ for $x_{N,1}\in S[c_1',r_1'']$.

Choose constants $r_N=(r_{N,1},\dots,r_{N,n})$, $x_{N,1}$ and K_N such that

$$K_N r_N^N\le 2^{-1}R',\quad K_N r_N^N\ge K_0,\quad r_N\le r'' ,$$

$$|App_N(F)-\textstyle\sum_J(-1)^{\#J+1}u(0_J)|\le 4^{-1}R' ,$$

$$N_1(r_{N,1}^{mx})^{p_1}r_N'^{p'}\le d(\sin 2e - \sin e) ,$$

$$C_N(1+(d\sin e)^{-1})+(4^{-1}+(d\sin e)^{-1}BK_Nr_N^N)K_N(\max\{1,r_N^N\}))\le K_N$$

$$(4^{-1}+2BK_Nr_N^N(d\sin e)^{-1})\le 2^{-1} .$$

Notice that we can choose $r_{N,i}=r_i''$ for $i=2,\dots,n$ if $N_1>0$. For $f\in V_N[m,c',r_N,t]$, set

$$(L_Nf)_j = U_j(x)\int_{r_{j,N}(x)} U_j(y_j(s),x')^{-1}x'^{-p'}y_j(s)^{-p_1-1}b_j^N(y_j(s),x':f)dy_j(s)$$

for $j\notin NI(c',U)$ and

$$(L_Nf)_j = U_j(x)U_j(x_{N,1},x')^{-1}(u_{0,j}(x_{N,1},x')-App_N(F)(x_{N,1},x')_j)$$

$$+U_j(x)\int_{r_{j,N}(x)}U_j(Y_j(s),x')^{-1}x'^{-p'}y_j(s)^{-p_1-1}b_j^N(y_j(s),x':f)dy_j(s)$$

for $j\in NI(c',U)$, where

$$(b_j^N)_{j\in[1,m]} = E_N(x)+F_N(x,u)-Tu$$

and the paths of integral $r_{j,N}(x)$ are defined as above with $t((p,o))$ and $x_{N,1}$. Then, by using Lemma 2.2 and by the choice of constants, we see that L_N is a contraction operator of $V_{N,K_N}[m,c',r_N,t]$. Therefore, there exists a unique $u_N\in V_{N,K_N}[m,c',r_N,t]$ satisfying $L_Nu_N=u_N$.

For $f\in V_0[m,c',r_N,t]$,

$$E_N(x)+F_N(x,f-App_N(F))-T(f-App_N(F)) = a(x,f)-Tf-(x^p(y\partial/\partial y)+D)App_N(F) .$$

As we see

$$x^py(\partial/\partial y)(U(y,x')^{-1}App_N(F)) = U(y,x')^{-1}(x^p(y\partial/\partial y)+D)App_N(F) ,$$

we obtain

$$U_j(x)\int_{r_{j,N}(x)}U_j(y_j(S),x')^{-1}((\partial/\partial y_j(s))+x'^{-p'}y_j(s)^{-p_1-1}d_j)(App_N(F)(y_j(s),x')_j)dy_j(s)$$

$$=\begin{cases} App_N(F)_j, \text{ for } j\notin NI(c',U) , \\ App_N(F)_j-U_j(x)U_j(x_{N,1},x')^{-1}App_N(F)(x_{N,1},x')_j, \text{ for } j\in NI(c',U) . \end{cases}$$

Therefore, we obtain

$$L_N(f-App_N(F))+App_N(F) = Lf ,$$

where Lf is defined by (*) and (**), because, for $j \notin NI(c',U)$, $r_{j,0}(x)=r_{j,N}(x)$ and for $j \in NI(c',U)$, $(Lu_0)_j(x_{N,1},x')=u_0(x_{N,1},x')_j$. Put $f_N=u_N+App_N(F)$. Then, we see that

$$L_N(f_N-App_N(F))+App_N(F) = f_N \text{ , i.e. } Lf_N = f_N .$$

The operator L is a contraction of $V_{0,K_0'}[m,c',r_N,t]$ for $K_0'=K_N r_N^N$ $(\geq K_0)$, because of the choice of constants. As $V_{0,K_0'}[m,c',r_N,t]\supset V_{0,K_0}[m,c',r'',t]$, we obtain $f_N=u_0$ in $S[c',r_N]$, from which by the same argument as in the proof of Lemma I.2.1, we conclude that u_0 is strictly strongly asymptotically developable in $S[c',r'',t]$. From the integral equation $Lu_0=u_0$, we see easily that u_0 is the solution to (E.3) with the given initial value. Q.E.D.

<u>REMARK 2.3</u>. In the case that $a(x,u)$ is <u>linear with respect to</u> u, i.e. $a(x,u)=a_0(x)+A(x)u$, in order to prove THEOREM 2.3, it suffices to choose r'' so that

$$(d \sin e)^{-1}|A| \leq 8^{-1} ,$$

$$N_1(r_1^{mx''})^{p_1}(r'')^{p'} \leq d(\sin 2e - \sin e) ,$$

$$C_N(1+(d \sin e)^{-1})+4^{-1}K_N \leq K_N .$$

<u>REMARK 2.4</u>. The sectorial domain $S[c',r'',t]$ includes a polysector $S(c',r''')$ for some r''' less than r''.

<u>REMARK 2.5</u>. By THEOREM of Borel-Ritt type (see THEOREM I.2.2 in Section I.2), <u>there exists a function</u> $v_h(x')$ <u>with the property stated in THEOREM 2.3</u>.

<u>THEOREM 2.4</u>. <u>If</u> $A_0=D+T$ <u>is invertible, then, for any closed proper domain</u> $c'=\Pi_{i=1}^n c_i'$ <u>in</u> c <u>with respect to</u> $|U(x)|$ and for r'' <u>adequately chosen, there exists a strictly consistent family</u> $F=\{u(q_J)\}$ <u>of formal solutions to</u> (E.3) <u>in</u> $S[c',r']$.

<u>Moreover, the statement in THEOREM 2.3 is valid for this family</u> F.

PROOF. If $F=\{u(q_J)\}$ is a family of formal solutions of (E.3), then for any non-empty subset J of $[1,n]$ and any $q_J \in \mathbb{N}^J$, $u(q_J)$ satisfies the equation $(E.3.q_J)$. We shall prove that these equation can be solved by an induction on the cardinal number $n'=\#J$ and the length $L=|q_J|$ of index q_J.

In the case $n'=n$, i.e. $J=[1,n]$, we put $u_{0_J}=0$, then $(E.q_J)$'s can be solved uniquely by an inductive way on $|q_J|$. Suppose that we obtain a family of functions

$$\{u(q_J;x_I);\ \#J=n'+1,\ldots,n,\ q_J \in \mathbb{N}^J,\ I=J^c\}\ .$$

such that for any q_J $(\#J=n'+1,\ldots,n)$, $u(q_J;x_I)$ is a solution of $(E.3.q_J)$, holomorphic and strictly strongly asymptotically developable in $\Pi_{i\in I}S[c_i',r_{J,i}]$, where $r_{J,i}$'s $(i\in I=J^c)$ are positive constant. We now proceed to the case n'. If $J\cap J(+p)\neq\emptyset$, then we can solve the equation $(E,0_J)$ by PROPOSITION 2.1. Let $u(0_J;x_I)$ be a solution holomorphic and strictly strongly asymptotically developable in the sector $\Pi_{i\in I}S[c_i',r_{0_J,i}]$. By taking r_J less than r_{0_J}, if necessary, we can assume that the matrix

$$(A_0-(\partial/\partial u)a(x,u)|_{u=u(0_J),x_J=0})$$

is invertible in $\Pi_{i\in I}S[c_i',r_{J,i}]$. Then, we can solve the equations $(E.3.q_J)$ inductively on $|q_J|$ in $\Pi_{i\in I}S[c_i',r_{J,i}]$.

If $\#J'=n'$ and $J'\cap J(+p)=\emptyset$, we apply THEOREM 2.3 to the equation $(E.q_{J'})$ and functions $v_{q_{J'},h}(x_{I'})$'s $(h\in NI(c'))$ such that the total family $TA(v_{q_J,h}(x_{I'}))$ of strictly strongly asymptotic expansion coincides with

$$\{u(q_{J'\cup J''};x_{1,0},x_{I''})_h;\ J''\subset[2,n]-J',\ J''\neq\emptyset,\ I''=I'-J'',\ q_{J''}\in\mathbb{N}^{J''}\}\ .$$

For $0_{J'}$, let $u(q_{J'})$ be a solution of $(E.0_{J'})$, holomorphic and strictly strongly asymptotically developable in $\Pi_{i\in I'}S[c_i',r_{0_{J'},i}]$. By taking r_J still smaller, if necessary, we have estimate

$$|(\partial/\partial u)a(x,u)|_{u=u(0_J),x_J=0}|\leq 8^{-1}\ ,$$

for $x_I\in\Pi_{i\in I}S[c',r_{J,i}]$. Then, the linear equation $(E.3.q_{J'})$ has a solution $u(q_{J'})$

holomorphic and strictly strongly asymptotically developable in the sector $\Pi_{i \in I} S[c_i', r_{J',i}]$. By the construction, we can verify the consistency of the family obtained and we see that it is a family of formal solutions in $S[c', r']$. Q.E.D.

SECTION II.3. EXISTENCE THEOREMS OF ASYMPTOTIC SOLUTIONS TO INTEGRABLE SYSTEMS.

In this section, we keep the notation used in the preceding section.

Let A_i, $i=1,\dots,n'\leqq n$, be m-by-m constant matrices and $b_i(x,u)$ be m dimensional column vectors of functions holomorphic and strongly asymptotically developable as x tends to $H'=\{(x,u): x_1\dots x_n=0\}$ in $S(c,r)\times D(R)$. By the holomorphy with respect to u, for any $i+1,\dots,n'$, b_i has the following expansion

$$b_i(x,u) = \sum_{q\in\mathbb{N}^m, q\neq 1} b_{i,q}(x)u^q + B_i(x)u ,$$

where $b_{i,q}(x)$'s ($|q|\neq 1$, $q\in\mathbb{N}^m$) (resp. $B_i(x)$'s) are m-dimensional column vectors (m-by-m matrices) of functions holomorphic and strongly asymptotically developable in $S(c,r)$. In the following, we suppose that for any $i=1,\dots,n'$

$$FA(b_{i,0})(0) = \lim_{x\to 0} b_{i,0}(x) = 0 \quad\text{and}\quad FA(B_i)(0) = \lim_{x\to 0} B_i(x) = 0 .$$

We shall consider systems of equations

$$\text{(S.1)} \quad A_i u = b_i(x,u) , \quad i=1,\dots,n' ,$$

with conditions

$$\text{(C.I.1)}_{i,j} \quad (A_i-(\partial/\partial u)b_i)(A_j u-b_j) = (A_j-(\partial/\partial u)b_j)(A_i u-b_i)$$

$i,j=1,\dots,n'$: we need this study to construct formal solutions to systems of differential equations.

PROPOSITION 3.1. <u>If</u> A_i <u>is invertible for some</u> $i=1,\dots,n'$, <u>then, for any closed subset</u> $c'=\Pi_{i=1}^n c_i'$ <u>in</u> c <u>and for some</u> r', <u>there exists a strictly consistent family</u> F <u>of formal solution to</u> (S.1) <u>with</u> PS(F)(0)=0, <u>and there exists a unique solution u to</u> (S.1) <u>which is holomorphic in</u> S[c',r'] <u>and strictly strongly asymptotically developable in</u> S[c',r'] <u>with</u> TA(u)=F.

PROOF. By PROPOSITION 2.2, for the ith equation

$$(S.1)_i \quad A_i u = b_i(x,u) ,$$

there exists such a family F and such a solution u. Then, the condition $(C.I.1)_{i,j}$ implies that

$$w_j = A_j u(x) - b_j(x,u(x))$$

satisfies the linear equation

$$(A_i - (\partial/\partial u) b_i(x,u(x)))w = 0 ,$$

for any $j \neq i$. We see easily that w_j's are holomorphic in $S[c',r']$ and strictly strongly asymptotically developable in $S[c',r']$ with $FA(w_j)(0)=0$. By PROPOSITION 2.2 again, w_j's are equal to zero. Q.E.D.

Corollary 3.1. *If* A_i *is invertible for some* $i=1,\ldots,n'$ *and if* $b_j(x,u)$ *is strongly asymptotically developable to an element in* $\mathcal{O}_{D(r)\times D(R)}\hat{|}_{H'}(D(r)\times D(R))^m$ *for any* $j=1,\ldots,n'$, *then there exists a unique formal power-series solution* u *in* $\mathcal{O}_{D[r']}\hat{|}_H(D[r'])^m$ *satisfying* $u(0)=0$ *and a unique solution* u *holomorphic and strongly asymptotically developable to* u *in* $S[c',r'']$, *where* $r'' \leq r' < r$ and $c' \subset c$.

Corollary 3.2. *If* A_i *is invertible for some* $i=1,\ldots,n'$ *and if* $b_j(x,u)$ *is holomorphic in* $D(r)\times D(R)$ *for any* $j=1,\ldots,n'$, *there exists a unique convergent power-series solution* u *satisfying* $u(0)=0$.

We now consider a system of differential equations of the first order of the form

$$(S) \qquad x^{p_i}(x_i\partial/\partial x_i)u + A_i u = b_i(x,u) , \quad i=1,\ldots,n' \leq n ,$$

under the conditions

$$(C.I)_{i,j} \quad \begin{cases} (x_i\partial/\partial x_i)(x^{-p_j}a_j) + x^{-p_j-p_i}((\partial/\partial u)a_j)a_i \\ = (x_j\partial/\partial x_j)(x^{-p_i}a_i) + x^{-p_i-p_j}((\partial/\partial u)a_i)a_j \end{cases}$$

for $i,j=1,\ldots,n'$, where $p_i=(p_{i,1},\ldots,p_{i,n}) \in \mathbb{N}^n$ and $a_i = b_i - A_i u$ for $i=1,\ldots,n'$. We can easily verify

LEMMA 3.1. *For a m-dimensional vector function* u *of* x, *the following equalities are valid*:

$$(x^{p_i}(x_i\partial/\partial x_i)-(\partial/\partial u)a_i(x,u(x))-p_{ji}x^{p_i})w_j$$
$$= (x^{p_j}(x_j\partial/\partial x_j)-(\partial/\partial u)a_j(x,u(x))-p_{ij}x^{p_j})w_i$$

for $i,j=1,\dots,n'$, *where* w_j *is defined as*

$$w_j = x^{p_j}(x_j\partial/\partial x_j)u(x)-a_j(x,u(x)) \ , \ j=1,\dots,n' \ .$$

Using THEOREMS 2.1, 2.2 (resp. THEOREMS 2.3, 2.4) and LEMMA 3.1, by the same argument as in the PROPOSITION 3.1, we can deduce THEOREMS 3.1 and 3.2 (resp. THEOREMS 3.3 and 3.4):

THEOREM 3.1. *Suppose that* $p_i=0$ *for some* $i\in[1,n']$ *and suppose that* $F=\{u(a_J)\}$ *is a strictly consistent family of formal solutions to (S) in* $S[c',r']$ *with* $PS(F)(0)=0$, *where* $c'=\Pi_{i=1}^n c_i'$ *is a closed subset in c and* r' *is a constant less than* r. *Then, (S) has a unique solution u in* $V_0[m,c',r'']$ *which is strictly strongly asymptotically developable in* $S[c',r'']$ *and whose total family of coefficients of asymptotic expansion coincides with the family F of formal solutions, where* r'' *is some constant adequately chosen*.

THEOREM 3.2. *If* $p_i=0$ *for some* $i\in[1,n']$ *and if* $A_i+(k-p_{ji})I_m$ *is invertible for any nonnegative integer k and for any* $j\in[1,n']$, *then there exists only one family* F *of formal solutions to (S) satisfying* $PS(F)(0)=0$ *in any closed subpolysector* $S[c',r']$ *in* $S(c,r)$ *with* r' *sufficiently small*.

For this family F, the statement in THEOREM 3.1 is valid.

Corollary 3.3. *Suppose that* $b_j(x,u)$ *is strongly asymptotically developable to an element in* $\mathcal{O}_{D(r)\times D(R)}\hat{|}_{H'}(D(r)\times D(R))^m$ *for any* $j=1,\dots,n'$ *and that* $p_i=0$ *for some* $i=1,\dots,n'$. *If* (S) *has a formal power-series solution* $\hat{u}$ *in* $\mathcal{O}_{D[r']}\hat{|}_H(D[r'])^m$

satisfying $\hat{u}(0)=0$, then there exists a unique solution u holomorphic and strongly asymptotically developable to $\hat{u}$ in $S[c',r'']$, where $r'' \leq r' < r$ and $c' \subset c$. Moreover if A_i+kI_m is invertible for any nonnegative integer and for the same i, then there exists a unique solution u holomorphic and strongly asymptotically developable to an element $\mathcal{O}_{D[r']}\hat{|}_H(D[r'])^m$ in $S[c,r'']$ satisfying $\lim_{x\to 0}u(x)=0$.

Corollary 3.4. Suppose that $b_j(x,u)$ is holomorphic in the disc $D(r)\times D(R)$ for any $j=1,\ldots,n'$ and that $p_i=0$ for some $i=1,\ldots,n'$. If (S) has a formal power-series solution u in $\mathcal{O}_{D[r']}\hat{|}_H(D[r'])^m$, then u is convergent. Moreover, if A_i+kI_m is invertible for any nonnegative integer and for the same i, then there exists a unique convergent power-series solution u to (S) with $u(0)=0$.

THEOREM 3.3. We assume that $p_{ii} > 0$ and $A_i=D_i+T_i$ is invertible for some $i \in [1,n']$. Let c' be a non-negatively strictly proper domain in c with respect to $|U^i(x)|=|\exp(p_{ii}{}^{-1}D_i x^{-p_i})|$, and suppose that $F=\{u(q_J)\}$ is a strictly consistent family of formal solutions to (S) in $S[c',r']$ with $PS(F)(0)=0$.

Then, there exists a unique solution u(x) to (S) in $V_0[m,c',r'',t]$, such that u is strictly strongly asymptotically developable in $S[c',r'']$ with $TA(u)=F$, where $0 < r'' < r'$.

THEOREM 3.4. Under the same assumptions as in THEOREM 3.3, there exists a strictly consistent family $F=\{u(q_J)\}$ of formal solutions to (S) in $S[c',r'']$, where c' is a non-negatively strictly proper domain with respect to $|U^i(x)|$ and r'' is a constant less than r'. Moreover, the statement in THEOREM 3.3 is valid for this family F.

Consider now the system (S) in the case that $p_{i,i} > 0$ for all $i \in [1,n']$. Then, by the conditions $(C.I)_{i,j}$'s, we see that

$$A_iA_j = A_jA_i \ , \quad i,j=1,\ldots,n' \ ,$$

and so we can suppose, without loss of generality, that A_i's are written in the form

$$A_i = D_i + T_i$$

with the diagonal matrices $D_i = \oplus_{k=1}^{m} d_{i,k}$, and the properly upper triangular matrices T_i, $i=1,\ldots,n'$. We write

$$U_{i,k}(x) = \exp(-p_{i,i} d_{i,k} x^{-p_i}) ,$$

for $i \in [1,n']$ and for $k \in [1,m]$, and write

$$U^i(x) = \oplus_{k=1}^{m} U_{i,k}(x) ,$$

for $i=1,\ldots,n'$. In the following, we use the notation

$$x_{[1,n']} = (x_1,\ldots,x_{n'}),\ x_{[n'+1,n]} = (x_{n'+1},\ldots,x_n) ,$$

$$x_{0,[1,n']} = (x_{0,1},\ldots,x_{0,n'}),\ x_{0,[n'+1,n]} = (x_{0,n'+1},\ldots,x_{0,n}) ,$$

$$p_{i,[1,n']} = (p_{i,1},\ldots,p_{i,n'}),\ p_{i,[n'+1,n]} = (p_{i,n'+1},\ldots,p_{i,n}) ,$$

$$x_{[n'',n]}^{p_{i,[n'',n]}} = \Pi_{j=n''}^{n} x_j^{p_{i,j}} ,$$

where $n'' \in [1,n]$.

We shall prove the following theorems:

THEOREM 3.5. *We assume that* $p_{i,i} > 0$ *and* $A_i = D_i + T_i$ *is invertible for all* $i \in [1,n']$. *Let* $c' = \Pi_{i=1}^{n} c_i'$ *be a strictly proper domain in c with respect to all* $|U^i(x)|$'s $(i=[1,n'])$, *and suppose that* $F = \{u(q_J)\}$ *is a strictly consistent family of formal solutions to (S) in* $S[c',r']$ *with* $PS(F)(0)=0$. *Let* $v_k(x_{[n'+1,n]})$'s $(k \in \cap_{i=1}^{n'} NI(c',U^i))$ *be functions such that, for any subset* J' *of* $[n'+1,n]$ *and* $q_{J'} \in \mathbb{N}^{J'}$,

$$TA(v_k)_{q_{J'}} = (u(q_{J'})(x_{0,[1,n']},x_{I'}))_k \quad (k \in \cap_{i=1}^{n'} NI(c',U^i)) ,$$

where $I' = J'^c$ *in* $[n'+1,n]$ *and* $x_{0,[1,n']}$ *is sufficiently small.*

Then, there exists a unique solution $u(x)$ *to (S) in* $V_0[m,c',r'']$, *such that* u *is strictly stongly asymptotically developable in* $S[c',r'']$ *with* $TA(u)=F$, *and that for* $k \in \cap_{i=1}^{n'} NI(c',U^i)$

$$u_k(x_{0,[1,n']},x_{[n'+1,n]}) = v_k(x_{[n'+1,n]}) ,$$

where $0 < r'' < r'$ *adequately chosen*.

THEOREM 3.6. *If* $A_i = D_i + T_i$ $(i=[1,n'])$ *are invertible, then, for any* $c' = \Pi_{i=1}^{n} c_i'$ *close proper domain in c with respect to all* $|U^i(x)|$, $i \in [1,n']$ *and for r' sufficiently small, there exists a strictly consistent family* $F = \{u(q_J)\}$ *of formal solutions to (S) in* $S[c',r']$ *with* $PS(F)(0)=0$. *Moreover, the statement in* THEOREM 3.5 *is valid for this family F of formal solutions.*

In order to prove THEOREM 3.5, we prepare two propositions.

PROPOSITION 3.2. *If* $u(x) = (u_k(x))_{k \in [1,m]}$ *is a solution to* (S) *with the properties stated in* THEOREM 3.5, *then* $u_k(x)$ *satisfies the following integral equation: if* $k \in \cap_{i=1}^{n'} NI(c',U^i)$, *then*

$$u_k(x) = \sum_{i=1}^{n'} U_{i,k}(x_{0,[1,i-1]},x_{[i,n]})$$

$$\times \int_0^2 W_{i,k}(x,u(x))|_{x_{[1,i-1]}=x_{0,[1,i-1]},x_i=y_{i,k}(s)} dy_{i,k}(s)$$

$$+U_{n',k}(x_{0,[1,n'-1]},x_{[n',n]}) U_{n',k}(x_{0,[1,n']},x_{[n'+1,n]}) v_k,$$

if $k \in \cap_{i=1}^{j-1} NI(c',U^i) - \cap_{i=1}^{j} NI(c',U^i)$, $j=2,\ldots,n'$, *then*

$$u_k(x) = \sum_{i=1}^{j-1} U_{i,k}(x_{0,[1,i-1]},x_{[i,n]})$$

$$\times \int_0^2 W_{i,k}(x,u(x))|_{x_{[1,i-1]}=x_{0,[1,i-1]},x_i=y_{i,k}(s)} dy_{i,k}(s)$$

$$+U_{j,k}(x_{0,[1,j-1]},x_{[j,n]})$$

$$\times \int_0^2 W_{j,k}(x,u(x))|_{x_{[1,j-1]}=x_{0,[1,j-1]},x_j=z_{j,k}(s)} dz_{j,k}(s) ,$$

otherwise,

$$u_k(x) = U_{1,k}(x)\int_0^2 W_{1,k}(x,u(x))|_{x_1=z_{1,k}(s)}dz_{1,k}(s) ,$$

where $W_{i,k}(x,u(x))$ are defined by

$$W_{i,k}(x) = U_{i,k}(x)^{-1}x^{-p_i}(a_{i,k}(x,u(x))-d_{i,k}u_k(x)) ,$$

and $(y_{i,k}(s))_{s\in[0,2]}$'s are respectively defined as paths jointing $(x_{0,[1,i]},x_{[i+1,n]})$ to x by the same way in the case where c' is a negatively strictly proper domain, and $(z_{i,k}(s))_{s\in[0,2]}$'s are respectively defined as paths jointing $(0,x_{[i+1,n]})$ to x by the same way as in the case where c' is a non-negatively strictly proper domain (see Section II.2).

PROOF. By integrating the equations inductively and by using the asymptotic properties of $U_{i,k}$'s, we arrive at the conclusion. Q.E.D.

PROPOSITION 3.3. The system of integral equations stated in PROPOSITION 3.2 has a unique solution $u(x)$ with the properties stated in THEOREM 3.5.

PROOF. By the same argument as in the proof of THEOREM 2.3, we can deduce the existence and the uniqueness of solution with the properties. Q.E.D.

We now prove the following proposition and simultaneously THEOREM 3.5 by an induction on n'.

PROPOSITION 3.4. The unique solution to the system of integral equations is a unique solution of (S) with the properties in THEOREM 3.5.

PROOF of PROPOSITION 3.4 and THEOREM 3.5. In the case $n'=1$, we proved the existence and the uniqueness of solution of integral equations in the proof of THEOREM 2.3 and THEOREM 2.3 itself implies THEOREM 3.5. We suppose that the statement of PROPOSITION 3.4 and THEOREM 3.5 are true in the case of the numbers less than n' and proceed to the case n'. First, we prove PROPOSITION 3,4 and then THEOREM 3.5 is evidently true by PROPOSITIONS 3.2, 3 and 4.

We regard the system of integral equations for the case n' as a system of integral equations for the case n'-1 with the initial functions

$$U_{n'-1,k}(x_{0,[1,n'-2]},x_{[n'-1,n]})^{-1}U_{n'-1,k}(x_{0,[1,n'-1]},x_{[n',n]})$$
$$\times U_{n',k}(x_{0,[1,n'-1]},x_{[n',n]})\int_0^2 W_{n',k}(x,u(x))|_{x_{0'[1,n'-1]},x_i=y_{n',k}(s)}dy_{n',k}(s)$$
$$+U_{n',k}(x_{0,[1,n']},x_{[n'+1,n]})v_k(x_{[n'+1,n]})$$

for $k\in\bigcap_{i=1}^{n'}NI(c',U^i)$, and

$$U_{n'-1,k}(x_{0,[1,n'-2]},x_{[n'-1,n]})^{-1}U_{n'-1,k}(x_{0.[1,n'-1]},x_{[n',n]})$$
$$\times U_{n',k}(x_{0,[1,n'-1]},x_{[n',n]})\int_0^2 W_{n',k}(x,u(x))|_{x_{0,[1,n'-1]},x_i=z_{n',k}(s)}dz_{n',k}(s)$$

for $k\in\bigcap_{i=1}^{n'-1}NI(c',U^i)-\bigcap_{i=1}^{n'}NI(c',U^i)$. Then, by the hypothesis of induction, u(x) is a solution of the system of equations

$$x^{p_i}(x_i\partial/\partial x_i)u=a_i(x,u(x)),\ i=1,\dots,n'-1\ .$$

Put

$$w_{n'}(x) = x^{p_{n'}}(x_{n'}\partial/\partial x_{n'})u(x)-a_{n'}(x,u(x))\ .$$

Then, by LEMMA 3.1, $w_{n'}(x)$ is a solution of the system of equations

$$x^{p_i}((x_i\partial/\partial x_i)-p_{n',i})-(\partial/\partial u)a_i(x,u(x))w_{n'}(x) = 0,\quad i\in[1,n'-1]\ ,$$

and $w_{n',k}(x_{0,[1,n'-1]},x_{[n',n]})=0$ for all $k\in\bigcap_{i=1}^{n'-1}NI(c',U^i)$. Thus, by the hypothesis of induction, in particular, by the uniqueness, we see that $w_{n'}=0$. Hence, we deduce that u is a solution of the system (S), and the uniqueness of solution of the system of integral equations implies that of the system (S). Q.E.D.

PROOF of THEOREM 3.6. We can prove THEOREM 3.6 by the same argument as in the proof of THEOREM 2.4.

Consider, for any non-empty subset J of [1,n], the formal completely integrable system $(S.3)_J=\{(E.3_{h,J}:h=1,\dots,n'\}$ of equations

$$(E.3)_{h,J} \quad x_I^{p_{h,I}} x_J^{p_{h,J}} (x_h\partial/\partial x_h)u_J + A_h u_J = b_{h,J}(x,u_J) \ ,$$

where $p_{h,I}=(p_{h,i})_{i\in I}$, $p_{h,J}=(p_{h,j})_{j\in J}$ and

$$b_{h,J} = FA_J(b_{h,0}) + FA_J(B_h)u_J + \sum_{q\in\mathbb{N}^n, |q|>1} FA_J(b_{i,q})u_J^q \ ,$$

satisfying the formal complete integrability condition

$$(C.I.3)_{i,h,J} \begin{cases} x^{p_h}((x_h\partial/\partial x_h)-p_{i,h})a_{i,J} + ((\partial/\partial u_J)a_{i,J})a_{h,J} \\ = x^{p_i}((x_i\partial/\partial x_i)-p_{h,i})a_{h,J} + ((\partial/\partial u_J)a_{h,J})a_{i,J} \end{cases}$$

for any $i,h=1,\dots,n'$, where $a_{h,J}=b_{h,J}-A_h u_J$. From the formal complete integrability condition, we can see that for any formal series $u_J=\sum_{q_J\in\mathbb{N}^J} u_{q_J}(x_I)x_J^{q_J}$, the following equalities are valid,

$$(E)_{i,h,J} \begin{cases} (x^{p_h}((x_h\partial/\partial x_h)-p_{i,h}) + A_h - (\partial/\partial u)b_{h,J}|_{u=u_J})w_{i,J} \\ = (x^{p_i}((x_i\partial/\partial x_i)-p_{h,i}) + A_i - (\partial/\partial u)b_{i,J}|_{u=u_J})w_{h,J} \end{cases}$$

for $i,h=1,\dots,n'$, where

$$w_{h,J} = x^{p_h}(x_h\partial/\partial x_h)u_J - a_{h,J}(x,u_J) \ , \quad h\in[1,n'] \ .$$

If a formal series $u_J=\sum_{q_J\in\mathbb{N}^J} u(q_J)x_J^{q_J}$ is a solution of $(S.3)_J$, then for any q_J, $u(q_J)$ satisties a system $(S.3)_{q_J}=\{(E.3.q_J)_h : h=1,\dots,n'\}$ of equations obtained from $(E.3.q_J)$ by replacing b_J with $b_{h,J}$ (see Section II.2). We shall prove that these systems can be solved inductively on the cardinal number $s=\#J$ and the length $L=|q_J|$

In the case $\#J=n$, i.e. $J=[1,n]$, the h-th equation $(E.3)_{h,J}$ has a unique solution $u_{[1,n]}$ under the condition $u_{0_{[1,n]}}=0$, and by the equalities $(E.3)_{h,i,J}$,

$$w_{i,[1,n]} = x^{p_i}(x_i\partial/\partial x_i)u_{[1,n]} - a_{i,J}(x,u_{[1,n]})$$

is a solution of the linear equation

$$(x^{p_h}((x_h\partial/\partial x_h)-p_{i,h})+A_h-(\partial/\partial u)b_{h,[1,n]}|_{u=u_{[1,n]}})w_{i,[1,n]}=0 ,$$

and $w_{i,[1,n]}|_{x=0}=0$. Hence, $w_{i,[1,n]}=0$, namely, $u_{[1,n]}$ is a unique solution of $(E.3)_{i,[1,n]}$ for all $i\in[1,n']$, and so $u_{[1,n]}$ is a unique solution of $(S.3)_{[1,n]}$. Suppose that we obtain a family of functions

$$\{u(q_J;x_I);\ \#J=s+1,\dots,n,\ q_J\in\mathbb{N}^J,\ I=J^c\}$$

such that for any q_J ($\#J=s+1,\dots,n$), $u(q_J;x_I)$ is a solution of $(S.3)_{q_J}$, holomorphic and strictly strongly asymptotically developable in $\Pi_{i\in I}S[c_i,r_{J,i}]$, where $r_J=(r_{J,1},\dots,r_{J,n})\in\mathbb{R}^+)^n$. We now proceed to the case $\#J=s$. Let J be a non-empty subset of $[1,n]$ such that $p_{h,J}>0$ for some $h\in[1,n']$. In this case, the h-th equation of the formal system $(E.3)_{h,J}$ has a unique solution $u_J=\sum_{q_J\in\mathbb{N}^J}u_{q_J}x_J^{q_J}$ by applying PROPOSITION 2.2 to the algebraic equation $(E.3.q_J)_h$ inductively on $|q_J|$. Then, by the same argument as above, we see that u_J is a unique solution of $(S.3)_J$.

Let J be a subset of $[1,n]$ such that $\#J=s$ and $p_{h,J}=0$ for all $h=1,\dots,n'$. Then, the equation $(E.3.q_J)_h$ is of the form

$$x^{p_h}(x_h\partial/\partial x_h)u_{q_J}+A_hu_{q_J}=(q_J!)^{-1}(\partial/\partial x_J)^{q_J}b_{h,J}(x,u_J)|_{x_J=0} ,$$

for all $h=1,\dots,n'$. We shall prove by an induction on $|q_J|$ that the system $(S.3)_{q_J}$ has a unique solution u_{q_J} in $\Pi_{i\in I}S[c_i,r_{J,i}]$ under the initial condition

$$u(q_J;x_{0,[1,n']},x_{I-[1,n']})_k = v_{q_J,k} \qquad (k\in\cap_{i=1}^{n'}NI(c',U^i))$$

where $v_{q_J,k}$ is a function of which the total family $TA(v_{q_J,k}(x_I))$ of strictly strongly asymptotic expansion coincides with

$$\{u(q_{J\cup J'};x_{0,[1,n']},x_{I'})_k;\ J'\subset[n'+1,n]-J,\ J'\neq\emptyset,\ I'=I-J',\ q_{J'}\in\mathbb{N}^{J'}\}$$

for all $k\in\cap_{i=1}^{n'}NI(c',U^i)$. In the case $|q_J|=0$, by putting $x_J=0$ in the equalities

$(C.I.3)_{i,h,J}$, $i,h\in[1,n']$, we can verify that the system $(S.3)_{0_J}$ is completely integrable, and so we can apply THEOREM 3.5 to the system $(S.3)_{0_J}$ and the initial functions $v_{0_J,k}$, $k\in\bigcap_{i=1}^{n'}NI(c',U^i)$. We assume that we obtain functions $u(q_J)$ $(|q_J|=0,\ldots,L-1)$. Then, for any q_J with $|q_J|=L$, by operating $(q_J!)^{-1}(\partial/\partial x_J)^{q_J}$ to each sides of $(C.I.3)_{i,h,J}$ and by putting $a_J=0$, we can see that the completely integrable condition of the system $(S.3)_{q_J}$ is satisfied and so we can apply THEOREM 3.5 to this system.

Thus, we obtain a family $\{u(q_J)\}$ of functions inductively. By the construction, we see that the family is a strictly consistent family of formal solution to (S). Q.E.D.

Finally, we give a remark on generalizations of THEOREM 3.1-6.

REMARK 3.1. Let s be a positive integer and let m_k, $k=1,\ldots,s$ be integers such that $m_1+\ldots+m_s=m$. Consider a system (S) of differential equations $(E.3)_h$, $h=1,\ldots,n'$, which are simultaneously decomposed in the form

$$(E.3)_{h,k}\quad x^{p_{h,k}}(x_h\partial/\partial x_h)v^k+A_{h,k}v^k=b_{h,k}(x,v^k)\ ,$$

$h=1,\ldots,n'$, $k=1,\ldots,s$, where $p_{h,k}\in\mathbb{N}^n$, $A_{h,k}$ is an m_k-by-m_k matrix and $b_{h,k}$ is m_k-dimensional column vector of functions holomorphic in the open set $S(c,r)\times D(R)^{m_k}$ in $\mathbb{C}^n\times\mathbb{C}^{m_k}$ and strongly asymptotically developable as x tends to H in $S(c,r)$ uniformly with respect to u^k in $D(R)^{m_k}$, for any $h=1,\ldots,n'$ and any $k=1,\ldots,s$. If, for any $k=1,\ldots,s$, each system $(S.3)_k$ of equations $(E.3)_{h,k}$ $(h=1,\ldots,n')$ satisfies the assumptions of THEOREMS 3.2, 3.4 or 3.6, then the system (S) has a consistent family of formal solutions and a solution with the properties stated in THEOREMS 3.2, 3.4 or 3.6. Hence, if there exists a transformation $u=p(x,v_1,\ldots,v_s)$ such that P is an invertible m-by-m matrix of functions holomorphic in $S(c,r)\times D(R)^m$ and strongly asymptotically developable with respect to x in $S(c,r)$ uniformly with respect to v in $D(R)^m$, and such that the transformed system with respect to v is decomposed system as above, then the former system (S) has a family of formal solutions and solution holomorphic and strongly asymptotically developable with the properties as above.

SECTION II.4. LIST OF THEOREMS OF EXISTENCE AND UNIQUENESS OF SOLUTIONS TO INTEGRABLE SYSTEMS.

Let $D(r)\times D(R)=\Pi_{i=1}^{n}D(r_i)\times\Pi_{j=1}^{m}D(R_j)$ be a open polydisc at the origin with radii $r=(r_1,\ldots,r_n)$, $R=(R_1,\ldots,R_m)$ in $\mathbb{C}^{n+m}$ with holomorphic coordinates $x_1,\ldots,x_n,u_1,\ldots,u_m$, and let H be the locus defined by the equation $x_1\ldots x_{n''}=0$ in $D(r)\times D(R)$.

Let A_i, $i=1,\ldots,n'\leq n$, be m-by-m constant matrices and $b_j(x,u)$ be m-dimensional column vectors of functions holomorphic in a polysector

$$S(c,r;R) = \Pi_{i=1}^{n''}S(c_i,r_i)\times\Pi_{i=n''+1}^{n}D(r_i)\times D(R) ,$$

and strongly asymptotically developable as x tends to H in S(c,r;R) uniformly with respect to $(x_{n''+1},\ldots,x_n,u)$. By the holomorphy with respect to u, for any $i=1,\ldots,n'$, b_i has the following expansion

$$b_i(x,u) =\sum_{q\in\mathbb{N}^n,\ |q|\neq 1} b_{i,q}(x)u^q+B_i(x)u ,$$

where $b_{i,q}(x)$'s ($|q|\neq 1$, $q\in\mathbb{N}^m$) (resp. $B_i(x)$'s) are m-dimensional column vectors (resp. m-by-m matrices) of holomorphic and strongly asymptotically developable in $S(c,r)=\Pi_{i=1}^{n''}S(c_i,r_i)\times\Pi_{i=n''+1}^{n}D(r_i)$ uniformly with respect to $x_{[n''+1,n]}=(x_{n''+1},\ldots,x_n)$ in $D(r_{[n''+1,n]})=\Pi_{i=n''+1}^{n}D(r_i)$. In the following, we suppose that for any $i=1,\ldots,n'$

$$FA(b_{i,0})(0) = \lim_{x\to 0}b_{i,0}(x) = 0 \quad\text{and}\quad FA(B_i)(0) = \lim_{x\to 0}B_i(x) = 0 ,$$

Consider systems of equations

$$\text{(S.1)}\quad A_iu = b_i(x,u) ,\quad i=1,\ldots,n' ,$$

with conditions

$$\text{(C.I.1)}_{i,j}\quad (A_i-(\partial/\partial u)b_i)(A_ju-b_j) = (A_j-(\partial/\partial u)b_j)(A_iu-b_i)$$

$i,j=1,\ldots,n'$, and a system of differential equations of the first order of the form

$$(S) \quad x^{p_i}(e_i\partial/\partial x_i)u+A_iu = b_i(x,u) \ , \quad i=1,\dots,n'\leq n \ ,$$

under the conditions

$$(C.I)_{i,j} \quad \begin{aligned} &(e_i\partial/\partial x_i)(x^{-p_j}a_j)+x^{-p_j-p_i}(\partial/\partial u)a_ja_i \\ &=(e_j\partial/\partial x_j)(x^{-p_i}a_i)+x^{-p_i-p_j}(\partial/\partial u)a_ia_j \ , \end{aligned}$$

for $i,j=1,\dots,n'$, where $e_i=x_i$ $(i\leq n'')$, $e_i=1$ $(i>n'')$, $p_i=(p_{i1},\dots,p_{in''},0,\dots,0)\in\mathbb{N}^{n''}$ and $a_i=b_i-A_iu$ for $i=1,\dots,n'$.

If u is a solution of (S.1) (resp. (S)), holomorphic and strongly asymptotically developable in a subpolysector $S(c',r')$ of $S(c,r)$, then for any non-empty subset J of $[1,n'']$, the formal series $u_J=FA_J(u)$ satisfies the following formal system $(S.1)_J$ (resp. (S))

$$(S.1)_J \quad A_iu_J = b_{i,J}(x,u_J) \ , \quad i=1,\dots,n' \ ,$$

$$(S)_J \quad x_J^{p_J}x_I^{p_I}(e_i\partial/\partial x_i)u_J+A_iu_J = b_{i,J}(x,u_J) \ , \quad i=1,\dots,n' \ ,$$

where $I=J^c$, $p=(p_J,p_I,0,\dots,0)$ and

$$(4.1)_J \quad b_{i.J}(x,u_J) = FA_J(b_{i,0})+FA_J(B_i)u_J+\sum_{q\in\mathbb{N}^m}FA_J(b_{i,q})(u_J)^q \ .$$

Therefore, we have equations of $u(q_J)=TA(u)_{q_J}$ $(J\neq\emptyset, \subset[1,n],\ q_J\in\mathbb{N}^J)$ of the form $(Ei.1.q_J)$ (resp. $(Ei)_J$), $i=1,\dots,n'$

$$(Ei.1.q_J) \quad A_iu(q_J) = (q_J!)^{-1}(\partial/\partial x_J)^{q_J}(b_{i,J}(x,u_J(x)))|_{x_J=0} \ ,$$

$$(Ei.q_J) \quad LHS_i(q_J) = (q_J!)^{-1}(\partial/\partial x_J)^{q_J}(b_{i,J}(x,u_J(x)))|_{x_J=0} \ ,$$

where $LHS_i(q_J)$ is defined according to p_i and J as follows:

if $p_i=0$ and $i \in J$, $LHS_i(q_J)=(A_0+q_1 I_m)u(q_J)$,

if $p_i=0$ and $i \notin J$, $LHS_i(q_J)=(e_i \partial/\partial x_i)u(q_J)+A_i u(q_J)$,

if $p_i > 0$ and $J(+p_i) \cap J \neq \emptyset$, $LHS_i(q_J)=A_i u(q_J)$,

and if $p_i > 0$ and $J(+p_i) \cap J=\emptyset$, $LHS_i(q_J)=x_I^{p_I}(x_i \partial/\partial x_i)u(q_J)+A_i u(q_J)$,

where $J(+p_i)=\{j \in [1,n]: p_{ij} > 0\}$. By calculation, we can verify that the right-hand side is equal to

$$TA(b_{i,0})_{0_J}+TA(B_i)_{0_J}u(0_J)+\sum_{q \in \mathbb{N}^m, |q|>1} TA(a_{i,q})u(0_J)^q, \text{ if } |q_J|=0 ,$$

$$((\partial/\partial u)(a_i(x,u))|_{u=u(0_J),x_J=0})u(q_J) + \text{terms determined by}$$

$$b_{i,0},\ B_i,\ b_{i,q}\ (|q|>1) \text{ and } u(s_J)(|s_J|<|q_J|), \text{ if } |q_J|>0 .$$

DEFINITION 4.1. Let

$$F = \{f(x_I;q_J);\ J \subset [1,n''],\ J \neq \emptyset,\ I=J^c,\ q_J \in \mathbb{N}^J\}$$

be a consistent family in $S(c',r')$. We say that the family F is a family of formal solutions of (S.1) (resp. (S)), if for any non-empty set J of $[1,n'']$, the power series

$$PS(F) = \sum_{q_J \in \mathbb{N}^J} f(x_I;q_J)x_J^{q_J}$$

satisfies the system obtained from $(S.1)_J$ (resp. $(S)_J$) by replacing u_J with $PS(F)_J$, namely, $f(x_I;q_J)$ satisfies the equation obtained from $(Ei.1.q_J)$ (resp. $(Ei.q_J)$) by replacing $u(q_J)$ with $f(x_I;q_J)$ for any $J \neq \emptyset$ and any $q_J \in \mathbb{N}^J$.

PROPOSITION 4.1. If A_i is invertible for some $i=1,\ldots,n''$ (not n') then, for any proper open subset $c'=\Pi_{i=1}^{n''}c_i'$ of c and for some $r'=(r_1',\ldots,r_n')$, there exists a consistent family F of formal solutions to (S.1) with $App_0(F)(0)=0$, and there exists a unique solution u to (S.1) which is holomorphic in $S[c',r']$ and strongly asymptotically developable in $S(c',r')$ with $TA(u)=F$.

THEOREM 4.1. Suppose that $p_i=0$ for some $i\in[1,n'']$ and suppose that $F=\{u(a_J)\}$ is a strictly consistent family of formal solutions of (S) in $S(c',r')$ with $PS(F)(0)=0$, where $c'=\prod_{i=1}^{n''}c_i'$ is a proper open subset of c and r' is a constant less than r. Then, (S) has a unique solution u holomorphic in $S(c',r'')$ which is strongly asymptotically developable in $S(c',r'')$ and whose total family of coefficients of asymptotic expansion coincides with the family F of formal solutions, where r'' is some constant adequately chosen.

THEOREM 4.2. If $p_i=0$ for some $i\in[1,n'']$ and if $A_i+(k-p_{ji})I_m$ is invertible for any nonnegative integer k and for any $j\in[1,n'']$, thenthere exists only one family F of formal solutions to (S) satisfying $PS(F)(0)=0$ in any proper open subpolysector $S(c',r')$ of $S(c,r)$ with r' sufficiently small.

For this family F, the statement in THEOREM 4.1 is valid.

DEFINITION 4.2. Let μ be a non-zero complex number. A subset set c' of $\mathbb{R}^{n''}$ or $T^{n''}$ is proper with respect to the function $|\exp(\mu x^{-p})| = \exp(|\mu|\cos(\arg\mu-(p,\arg x))|x|^{-p})$, if the set

$$c'\cap\{o\in\mathbb{R}^{n''}(\text{or } T^{n''});\ \cos(\arg\mu-(p,o))>0\}$$

has at most one connected component, where $(p,o)=\sum_{i=1}^{n''}p_i o_i$.

DEFINITION 4.3. A connected subset c' of $\mathbb{R}^{n''}$ (or $T^{n''}$) is strictly proper with respect to $|\exp(\mu x^{-p})|$, if there exists a positive constant e such that the set

$$c'\cap\{o\in\mathbb{R}^{N''}\ (\text{or } T^{n''});\ \cos(\arg\mu-(p,o))>-\sin 4e\}$$

has at most one connected component, namely, for any $o\in c'$

$$4e-3\pi/2<\arg\mu-(p,o)+2k\pi<-4e+3\pi/2$$

with some integer k.

If c' is sufficiently small, then this condition is satisfied.

DEFINITION 4.4. Suppose that $p=(p_1,\ldots,p_{n''},0,\ldots,0)\in\mathbb{N}^n-\{0\}$ and $d_1,\ldots,d_m$ are non-zero, and put $U_j(x)=\exp(d_jx^{-p})$, $j=1,\ldots,m$, $U(x)=\oplus_{j=1}^m U_j(x)$. A connected subset c' of $\mathbb{R}^{n''}$ (or $T^{n''}$) is said to be proper (resp. strictly proper) with respect to $|U(x)|$ if c' is proper (resp. strictly proper) with respect to $|U_j(x)|$ for all $j=1,\ldots,m$.

Let c' be a strictly proper domain with respect to $|\exp(\mu x^{-p})|$, and put

$$w_- = \inf\{(p,o);\ o\in c'\},\quad w_+ = \sup\{(p,o);\ o\in c'\}\ ,$$

$$w_r = \arg\mu+\pi/2 \text{ and } w_l = \arg\mu-\pi/2\ .$$

Then, there exists a positive constant d such that one of the following inequalities is valid modulo $\mathbb{Z}$ for all $o\in c'$:

(c-) $\quad w_l-\pi+4e \le w_- \le (p,o) \le w_r-4e$,

(c+) $\quad w_l+2d \le w_- \le (p,o) \le w_+ \le w_r-2d$,

(c-+) $\quad w_l-\pi+4e \le w_- \le w_l-2d \le (p,o) \le w_r-2d \le w_+ \le w_r$,

(c+-) $\quad w_l \le w_- \le w_l+2d \le (p,o) \le w_r+2d \le w_+ \le w_r-4e$,

or (c-+-) $\quad w_l-\pi+4e \le w_- \le w_l-2d \le (p,o) \le w_r+2d \le w_+ \le w_r+\pi-4e$,

where e and d are some positive constants.

DEFINITION 4.5. We say that c' is a negatively strictly proper briefly negative domain with respect to $|\exp(\mu x^{-p})|$ if all elements in c' satisfy the inequality (c-). In the other case, we say that c' in non-negatively strictly proper, briefly non-negative domain with respect to $|\exp(\mu x^{-p})|$.

Notice that $|\exp(\mu x^{-p})|$ is strictly strongly asymptotically developable to 0 in S(c',r') if c' is negatively strictly proper domain with respect to $|\exp(\mu x^{-p})|$.

If c' is a strictly proper domain with respect to $|U(x)|$, then we denote by NI(c,U) the subset of all j such that c' is a negatively strictly proper domain with respect to $|U_j(x)|$.

THEOREM 4.3. We assume that $p_{i,i}>0$ and $A_i=D_i+T_i$ is invertible for some $i\in[1,n']$. Let c' be a non-negatively strictly proper domain in c with respect to $|U^i(x)|=|\exp(p_{i,i}^{-1}D_i x^{-p_i})|$, and suppose that $F=\{u(q_J)\}$ is a consistent family of formal solutions to (S) in $S(c',r')$ with $PS(F)(0)=0$.

Then, there exists a unique solution $u(x)$ to (S), holomorphic in $S(c',r'')$, such that u is strongly asymptotically developable in $S(c',r'')$ with $TA(u)=F$, where $0<r''\leq r'$.

THEOREM 4.4. Under the same assumptions as in THEOREM 4.3, there exists a consistent family $F=\{u(q_J)\}$ of formal solutions to (S) in $S(c',r'')$, where c' is a non-negatively strictly proper domain with respect to $|U^i(x)|$ and r'' is a constant less than r'. Moreover, the statement in THEOREM 4.3 is valid for this family F.

Consider now the system (S) in the case that $p_{i,i}>0$ for all $i\in[1,n']$. Then, by the conditions $(C.I)_{i,j}$'s, we see that

$$A_iA_j = A_jA_i \ , \quad i,j=1,\ldots,n'' \ ,$$

and so we can suppose, without loss of generality, that A_i's are written in the form

$$A_i = D_i+T_i$$

with the diagonal matrices $D_i=\oplus_{k=1}^{m}d_{i,k}$, and the properly upper triangular matrices T_i, $i=1,\ldots,n'$. We write

$$U_{i,k}(x) = \exp(-p_{i,i}d_{i,k}x^{-p_i}) \ ,$$

for $i\in[1,n'']$ and for $k\in[1,m]$, and write

$$U^i(x) = \oplus_{k=1}^{m}U_{i,k}(x) \ ,$$

for $i=1,\ldots,n''$. In the following, we use the notation

$$n''' = \min\{ n'', n'\},$$

$$x_{[1,n''']} = (x_1,\dots,x_{n'''}), \quad x_{[n'''+1,n]} = (x_{n'''+1},\dots,x_n) ,$$

$$x_{0,[1,n''']} = (x_{0,1},\dots,x_{0,n'''}), \quad x_{0,[n'''+1,n]} = (x_{0,n'''+1},\dots,x_{0,n}) ,$$

THEOREM 4.5. We assume that $p_{i,i} > 0$ and $A_i = D_i + T_i$ is invertible for all $i \in [1,n''']$. Let $c' = \Pi_{i=1}^{n''} c_i'$ be a strictly proper domain in c with respect to all $|U^i(x)|$'s $(i=[1,n'''])$, and suppose that $F=\{u(q_J)\}$ is a consistent family of formal solutions of (S) in $S(c',r')$ with $PS(F)(0)=0$. Let $v_k(x_{[n'''+1,n]})$'s $(k \in \cap_{i=1}^{n'''} NI(c',U^i))$ be functions such that, for any subset J' of $[n'''+1,n'']$ and $q_{J'} \in \mathbb{N}^{J'}$,

$$TA(v_k)_{q_{J'}} = \{(u(q_{J'})(x_{0,[1,n''']},x_{I'}))_k\} \quad (k \in \cap_{i=1}^{n'''} NI(c',U^i)) ,$$

where $I'=J'^c$ in $[n'''+1,n'']$ and $x_{0,[1,n''']}$ is sufficiently small.

Then, there exists a unique solution $u(x)$ to (S), holomorphic in $S(c',r'')$, such that u is strongly asymptotically developable in $S(c',r'')$ with $TA(u)=F$, and that for $k \in \cap_{i=1}^{n'''} NI(c',U^i)$

$$u_k(x_{0,[1,n''']},x_{[n'''+1,n]}) = v_k(x_{[n'''+1,n]}) ,$$

where $0 < r'' < r'$ adequately chosen.

THEOREM 4.6. If $p_{i,i} > 0$ and $A_i = D_i + T_i$ $(i=[1,n'''])$ are invertible, then, for any $c' = \Pi_{i=1}^{n''} c_i'$ open proper domain in c with respect to all $|U^i(x)|$, $i \in [1,n''']$ and for r' sufficiently small, there exists a consistent family $F=\{u(q_J)\}$ of formal solutions to (S) in $S(c',r')$ with $PS(F)(0)=0$.

Moreover, the statement in THEOREM 4.5 is valid for this family F of formal solutions.

SECTION II.5. SPLITTING LEMMAS.

Following the idea of Y. Sibuya, we shall establish splitting theorems in the category of functions strongly asymptotically developable.

Let n'' be a positive integer inferier or equal to n and let H be a locus $\cup_{i=1}^{n''}\{x\in\mathbb{C}^n\colon x_i=0\}$ in $\mathbb{C}^n$. For an open subset c of $\mathbb{R}^{n''}$ and an n-tuple $r=(r_1,\dots,r_n)$ of positive real numbers, denote by A(c,r) the set of all functions holomorphic and strongly asymptotically developable in the open polysectorial domain

$$S(c,r) = \{x\in\mathbb{C}^n\colon (\arg x_1,\dots,\arg x_{n''})\in c,\ 0<|x_i|<r_i,\ i\leq n'',\ |x_i|<r_i,\ i>n''\}$$

Let n' be a positive integer equal or inferier to n and let be given m-by-m matrices $A_i(x)$, $i=1,\dots,n'$, of which elements belong to A(c,r). Suppose that $A_i(x)$ is commutative with the other matrices A_j $(j\neq i)$ for any $i=1,\dots,n''$. Denote by A_{i0} the limit of $A_i(x)$ as x tends to 0 in S(c,r) for any $i=1,\dots,n'$. Then, we see that

$$A_{i0}A_{j0} = A_{j0}A_{i0} \quad \text{for any} \quad i,j=1,\dots,n' ,$$

and so we can assume that each of A_{i0}'s is written in the following form:

$$(5.1)\qquad A_{i0} = \oplus_{k=1}^{s}(d_{ik}I_{m_k}+N_{ik}) = \oplus_{k_i=1}^{s_i}(d_{ik_i}I_{m_{k_i}}+N_{ik_i}) ,$$

where s_i and s are positive integers, m_{k_i}'s (resp. m_k's) are positive integers such that

$$\textstyle\sum_{k_i=1}^{s_i} m_{k_i} = m \ (\text{resp. } \sum_{k=1}^{s} m_k = m) ,$$

$I_{m_{k_i}}$ (resp. I_{m_k}) denotes the identity matrix of degree m_{k_i} (resp. m_k), N_{ik_i} (resp. N_{ik}) is a properly upper triangular matrix of degree m_{k_i} (resp. m_k) for $k_i=1,\dots,s_i$ (resp. $k=1,\dots,s$) and for each $i=1,\dots,n'$, d_{ik_i}'s are complex numbers such that

$$d_{ik_i}\neq d_{ik_i'},\ \text{if } k_i\neq k_i' ,$$

and d_{ik}'s ($i=1,\ldots,n'$, $k=1,\ldots,s$) are complex numbers such that

$$d_{ik} \neq d_{ik'} \text{ for some } i=1,\ldots,n' \text{ , if } k \neq k' \text{ .}$$

PROPOSITION 5.1. *According to the decomposition forms* (5.1) *of* A_{i0}'s, $A_i(x)$'s *are simultaneously decomposable: for any proper subset* c' *of* c *and for some* $r' \leq r$, *there exists an* m - *by* - m *invertible matrix* P *of functions in* $\mathcal{A}(c',r')$ *such that*

$$\lim_{x\to 0} P(x) = I_m \text{ ,}$$

$$P^{-1}(x)A_i(x)P(x) = \oplus_{k=1}^{s} B_{ik}(x) \text{ ,}$$

where $B_{ik}(x)$ *is an* m_k *by* m_k *matrix of functions in* $\mathcal{A}(c',r')$ *with*

$$\lim_{x\to 0} B_{ik} = d_{ik} I_{m_k} + N_{ik} \text{ ,}$$

or all $i=1,\ldots,n'$.

In order to prove this proposition, it suffices to prove the following lemma:

LEMMA 5.1. *For each* $i=1,\ldots,n'$, *according to the decomposition form* (5.1) *of* A_{i0}, $A_j(x)$'s ($j=1,\ldots,n'$) *are simultaneously decomposable*: *for any proper subset* c' *of* c *and for some* $r' \leq r$, *there exists an* m- *by* - m *invertible matrix* P_i *of functions in* $\mathcal{A}(c',r')$ *such that*

$$\lim_{x\to 0} P_i(x) = I_m \text{ ,}$$

$$P_i^{-1}(x)A_j(x)P_i(x) = \oplus_{k_i=1}^{s_i} B_{jk_i}(x) \text{ ,}$$

where $B_{jk_i}(x)$ *is an* m_{k_i} *by* m_{k_i} *matrix of functions in* $\mathcal{A}(c',r')$ *with*

$$\lim_{x\to 0} \oplus_{k_i=1}^{s_i} B_{jk_i} = A_{j0} \text{ ,}$$

for all $j=1,\ldots,n'$.

PROOF OF LEMMA 5.1. Let $P_i(x)$ be such a matrix of the form

$$P_i(x) = I_m-(P_{k_ik_i'}(x))_{k_i,k_i'=1,\dots,s_i} ,$$

where $P_{k_ik_i'}$ is m_{k_i} by $m_{k_i'}$ matricial function for any k_i, $k_i'=1,\dots,s_i$ and $P_{k_ik_i}=0$ for all $k=1,\dots,s$. Then, $P_{k_ik_i'}$'s satisfy the following system of equations

$$(A_{jk_ik_i'})(I_m-(P_{k_ik_i'})) = \oplus_{k_i=1}^{s_i} B_{jk}$$

where A_j is written in the blocknized form according to that of A_{i0} for all $j=1,\dots,n'$, that is,

$$A_{jk_ik_i} - \sum_{k_i''=1,k_i''\neq k_i}^{s_i} A_{jk_ik_i''}P_{k_i''k_i} = - B_{jk_i} ,$$

and

$$A_{jk_ik_i'} - \sum_{k_i''=1,k_i''\neq k_i'}^{s_i} A_{jk_ik_i''}P_{k_i''k_i'} = - B_{jk_i}P_{k_ik_i'} , \quad k_i \neq k_i' ,$$

from which we obtain

$$A_{jk_ik_i'} - \sum_{k_i''=1,k_i''\neq k_i'}^{s_i} A_{jk_ik_i''}P_{k_i''k_i'}$$

$$= -(A_{jk_ik_i} - \sum_{k_i''=1,k_i''\neq k_i}^{s_i} A_{jk_ik_i''}P_{k_i''k_i})P_{k_ik_i'} ,$$

for k_i, $k_i'=1,\dots,s_i$, $k_i\neq k_i'$, this system is rewritten in the following form of system of equations with respect to $Q=(Q_h)_{h\in[1,M]}$

$$T_j(x)Q = F_j(x)+ \sum_{h,h'=1}^{M} G_{jhh'}(x)Q_hQ_{h'} , \quad j=1,\dots,n' ,$$

where T_j's are M-by-M matrices of functions in A(c,r) for $j=1,\dots,n$, in particular

$$\lim_{x\to 0}T_i(x) = \prod_{k_i\neq k_i}(d_{ik_i}-d_{ik_i'})^{m_{k_i}m_{k_i'}} \neq 0 ,$$

$F_j(x)$'s are M-column vectors of functions in $\mathcal{A}(c,r)$ with

$$\lim_{x\to 0} F_j(x) = 0 \;,\quad j=1,\dots,n',$$

$G_{jhh'}(x)$'s are M-column vectors of functions in $\mathcal{A}(c,r)$ for $j=1,\dots,n'$, h, $h'=1,\dots,M$, and the following equalities are valid,

$$(T_j-(G_{jhh'}Q_{h'})_{h'=1,\dots,M})(T_{j'}-(G_{j'hh'}Q_{h'})_{h'=1,\dots,M})$$

$$= (T_{j'}-(G_{j'hh'}Q_{h'})_{h'=1,\dots,M})(T_j-(G_{jhh'}Q_{h'})_{h'=1,\dots,M}) \;,$$

for any j, $j'=1,\dots,n'$. Hence, by applying PROPOSITION 4.1. to this system, we reach the conclusion. Q.E.D.

Let be given differential operators ∇_i, $i=1,\dots,n'$, of the form

$$\nabla_i = e_i\partial/\partial x_i - x^{-p_i}A_i(x) \;,\quad i=1,\dots,n',$$

where $e_i=x_i$ $(i\leq n'')$, $e_i=1$ $(n''<i\leq n')$, $p_i=(p_{i1},\dots,p_{in''},0,\dots,0)\in\mathbb{N}^n$, $A_i(x)$ is an m-by-m matrix of functions in A(c,r). Suppose that ∇_i is commutative with the other operators ∇_j $(j\neq i)$ for any $i=1,\dots,n'$, i.e.

$$(e_i\partial/\partial x_i)(x^{-p_j}A_j(x))+x^{-p_i-p_j}A_j(x)A_i(x)$$

$$= (e_j\partial/\partial x_j)(x^{-p_i}A_i(x))+x^{-p_i-p_j}A_i(x)A_j(x) \;,$$

for i, $j=1,\dots,n'$.

<u>PROPOSITION 5.2.</u> <u>If, for</u> $i>n''$, $p_i=0$, <u>then there exists an invertible m-by-m matrix of holomorphic and strongly asymptotically developable in</u> S(c,r) <u>such that</u>

$$P^{-1}\cdot\nabla_i\cdot P = (\partial/\partial x_i), \text{ i.e. } P^{-1}\{A_iP-(\partial/\partial x_i)P\} = 0 \;,$$

$$P^{-1}\cdot\nabla_j\cdot P = (e_j\partial/\partial x_j)-x^{-p_i}B_i(x_1,\dots,x_{i-1},x_{i+1},\dots,x_n) \;,\quad j\neq i \;,$$

i.e. $P^{-1}\{x^{-p_j}A_jP-(e_j\partial/\partial x_j)P\} = x^{-p_j}B_j(x_1,\ldots,x_{i-1},x_{i+1},\ldots,x_n)$,

<u>where</u> B_j's <u>are m-by-m matrices of functions holomorphic and strongly asymptotically developable in</u> $\Pi^{n''}_{j=1,j\neq i}S(c_i,r_i)\times\Pi^{n}_{j=n''+1}D(r_j)$.

PROOF. We can easily see that the equation $(\partial/\partial x_i)u=A_iu$ can be solved in the category of functions holomorphic and strongly asymptotically developable (cf. THEOREM 4.2). Take P as a fundamental matrix of solutions of the equation $(\partial/\partial x_i)u=A_iu$. Then, evidently $P^{-1}\cdot\nabla_i\cdot P=(\partial/\partial x_i)$. Put $B_j=P^{-1}\{x^{-p_j}A_jP-(e_j\partial/\partial x_j)P\}$ for $j\neq i$, then, by the commutativity, we obtain $(\partial/\partial x_i)B_j=0$. Q.E.D.

Thus, we consider in the following under the condition that $P_i\neq 0$ for $i>n''$.

Denote by A_{i0} the limit of $A_i(x)$ as x tends to 0 in $S(c,r)$ for any $i=1,\ldots,n'$. Then, we see that

(5.2) if $p_i=0$ and $p_j=0$ for $i,j=1,\ldots,n''$, or if $p_i\neq 0$ and $p_j\neq 0$ for $i,j=1,\ldots,n'$,

$$A_{j0}A_{i0} = A_{i0}A_{j0} ,$$

(5.3) if $p_i=0$ and $p_j\neq 0$ for $i=1,\ldots,n''$, $j=1,\ldots,n'$,

$$-p_{ji}A_{j0}+A_{j0}A_{i0} = A_{i0}A_{j0} .$$

And so, we can assume that A_{i0}'s are written in the following forms:

$$(5.4)\quad A_{i0} = \oplus_{k=1}^{s}(d_{ik}I_{m_k}+N_{ik}) ,$$

for $i=1,\ldots,n'$, and if $p_i\neq 0$,

$$(5.5)\quad A_{i0} = \oplus_{k_i=1}^{s_i}(d_{ik_i}I_{m_{k_i}}+N_{ik_i}) ,$$

and if $p_i=0$,

$$(5.6)\quad A_{i0} = \oplus_{k_i=1}^{s_i}\oplus_{h_i=1}^{t_i}(d_{ik_ih_i}I_{m_{k_ih_i}}+N_{ik_ih_i}) ,$$

where s_i's, t_i's and s are positive integers, m_{k_i}'s, $m_{k_ih_i}$'s and m_k's are positve integers such that

$$\sum_{k_i=1}^{s_i} m_{k_i} = m \,, \quad \text{if} \quad p_i \neq 0 \,,$$

$$\sum_{k_i=1}^{s_i} \sum_{h_i=1}^{t_i} = m \quad (\text{and put } m_{k_i} = \sum_{h_i=1}^{t_i} m_{k_i h_i}) \,, \quad \text{if} \quad p_i = 0 \,,$$

$I_{m_{k_i}}$, $I_{m_{k_i h_i}}$ and I_{m_k} denote the identity matrices of degree m_{k_i}, $m_{k_i h_i}$, and m_k respectively, N_{ik_i}, $N_{ik_i h_i}$ and N_{ik} are properly upper triangular matrices of degree m_{k_i}, $m_{k_i h_i}$ and m_k respectively, d_{ik_i}'s, $d_{ik_i h_i}$'s are complex numbers such that

$$d_{ik_i} \neq d_{ik_i'} \,, \quad \text{if} \quad p_i \neq 0 \quad \text{and} \quad k_i \neq k_i' \,,$$

$$d_{ik_i h_i} \neq d_{ik_i' h_i'} \,, \quad \text{if} \quad p_i = 0 \quad \text{and} \quad k_i \neq k_i' \quad \text{or} \quad h_i \neq h_i' \,,$$

$$d_{ik_i h_i} \neq d_{ik_i' h_i'} + \mu \quad \text{for any integer } \mu, \quad \text{if} \quad p_i = 0 \quad \text{and} \quad k_i \neq k_i' \,,$$

$$d_{ik_i h_i} = d_{ik_i h_i'} + \mu_{h_i h_i'} \quad \text{for some integer, if } p_i = 0 \,,$$

with the decreasing series $\mu_{h_i h_i'} > \mu_{h_i h_i''}$ for $h_i' < h_i''$, and d_{ik}'s $(i=1,\ldots,n',$ $k=1,\ldots,s)$ are complex numbers such that if $k \neq k'$, then

$$d_{ik} \neq d_{ik'} \quad \text{for some } i=1,\ldots,n' \text{ with } p_i \neq 0,$$

or $d_{ik} \neq d_{ik'} + z$ for any integer z and for some $i=1,\ldots,n'$ with $p_i=0$.

<u>LEMMA 5.2</u>. <u>For each</u> $i=1,\ldots,n'$, <u>according to the decomposition form</u> (5.5) <u>or</u> (5.6) <u>of</u> A_{i0}, $A_j(x)$'s $(j=1,\ldots,n')$ <u>are simultaneously decomposable</u>:

(5.7) <u>if</u> $p_{ii} > 0$, <u>then, for non-negatively strictly proper subset</u> c' <u>of</u> c <u>with respect to all</u> $\exp((d_{ik}-d_{ik'})x^{-p_i})$, $k,k'=1,\ldots,s_i$, <u>and for some</u> $r' \leq r$,

or (5.8) <u>if</u> $p_i=0$, <u>then, for any proper subset</u> c' <u>of</u> c <u>and for some</u> $r' \leq r$, <u>there exists an m-by-m invertible matrix</u> P_i <u>of functions in</u> $\mathcal{A}(c',r')$ <u>such that</u>

$$\lim_{x \to 0} P_i(x) = I_m \,,$$

$$P_i^{-1}(x)\{A_j(x)P_i(x) - x^{p_j}(e_j \partial/\partial x_j)P_i(x)\} = \bigoplus_{k_i=1}^{s_i} B_{jk_i}(x) \,,$$

i.e. $P_i^{-1}(x)\cdot\nabla_j\cdot P_i(x) = \oplus_{k_i=1}^{s_i}((e_j\partial/\partial x_j)-x^{-p_j}B_{jk_i}(x))$,

<u>where</u> $B_{jk_i}(x)$ <u>is an</u> m_{k_i} <u>by</u> m_{k_i} <u>matrix of functions in</u> $\mathcal{A}(c',r')$ <u>with</u>

$$\lim_{x\to 0}\oplus_{k_i=1}^{s_i}B_{jk_i} = A_{j0} ,$$

<u>for all</u> $j=1,\ldots,n'$.

PROOF OF LEMMA 5.2. Let $P_i(x)$ be such a matrix of the form

$$P_i(x) = I_m-(P_{k_ik_i'}(x))_{k_i,k_i'=1,\ldots,s_i} ,$$

where $P_{k_ik_i'}$ is m_{k_i} by $m_{k_i'}$ matricial function for any k_i, $k_i'=1,\ldots,s_i$ and $P_{k_ik_i}=0$ for all $k=1,\ldots,s$. Then, $P_{k_ik_i'}$'s satisfy the following system of equations

$$x^{p_j}(e_j\partial/\partial x_j)(P_{k_ik_i'})+(A_{jk_ik_i'})(I_m-(P_{k_ik_i'}))$$

$$= (\oplus_{k_i=1}^{s_i}B_{jk_i})(I_m-(P_{k_ik_i'}))$$

where A_j is written in the blocknized form according to that of A_{i0} for all $j=1,\ldots,n'$, that is,

$$A_{jk_ik_i}-\sum_{k_i'=1,k_i'\neq k_i}^{s_i}A_{jk_ik_i'}P_{k_i'k_i} = B_{jk_i} ,$$

$$x^{p_i}(e_j\partial/\partial x_j)P_{k_ik_i'}+A_{jk_ik_i'}-\sum_{k_i''=1,k_i''\neq k_i'}^{s_i}A_{jk_ik_i''}P_{k_i''k_i'}$$

$$= -B_{jk_i}P_{k_ik_i'} , \quad k_i\neq k_i' ,$$

from which we obtain

$$x^{p_i}(e_j\partial/\partial x_j)P_{k_ik_i'}+A_{jk_ik_i'}-\sum_{k_i''=1,k_i''\neq k_i'}^{s_i}A_{jk_ik_i''}P_{k_i''k_i'}$$

$$= -(A_{jk_ik_i}-\sum_{k_i''=1,k_i''\neq k_i}^{s_i}A_{jk_ik_i''}P_{k_i''k_i})P_{k_ik_i'} ,$$

for k_i, $k_i'=1,\ldots,s_i$, $k_i\neq k_i'$, this system is completely integrable and rewritten in the following form

$$x^{p_J}(e_j\partial/\partial x_j)Q+T_j(x)Q=F_j(x)+\sum_{h,h'=1}^{M}G_{jhh'}(x)Q_hQ_{h'}\ ,\quad j=1,\ldots,n'\ ,$$

where T_j's are m-by-m matrices of functions in $\mathcal{A}(c,r)$ and

$$\lim_{x\to 0}T_j(x) = D_j+N_j\ ,$$

with N_j is a properly upper triangular matrix and D_j is a diagonal matrix similar to

$$\bigoplus_{k_j,k_j'=1,\ k_j\neq k_j'}^{s_j}(d_{jk_j}-d_{jk_j'})I_{m_{k_j}m_{k_j'}}\ ,\quad \text{if}\quad p_j\neq 0\ ,$$

or $$\bigoplus_{k_j,k_j'=1,\ k_j\neq k_j'}^{s_j}((\bigoplus_{h_j=1}^{t_j}d_{jk_jh_j}I_{m_{k_jh_j}})\otimes I_{m_{k_j'}}\ ,\quad \text{if}\quad p_j=0\ ,$$

for $j=1,\ldots,n'$, $F_j(x)$'s are column vectors of functions in $\mathcal{A}(c,r)$ with $\lim_{x\to 0}F_j(x)=0$, $j=1,\ldots,n'$, and $G_{jhh'}(x)$'s are column vectors of functions in $\mathcal{A}(c,r)$. Hence, by applying THEOREMS 4.4 or 4.2 to this system, we reach the conclusion. Q.E.D.

REMARK 5.1. If $A_i(x)$'s $(i=1,\ldots,n)$ are holomorphic in a polydisc $D(r)=\Pi_{i=1}^{n}D(r_i)$, then, by the construction, $App_{p_i}(P_{k_ik_i'};x)$'s $(k_i,k_i'=1,\ldots,s_i)$ are convergent series, i.e. holomorphic in $D(r)$ and so $App_{p_i}(B_{jk_i};x)$'s $(j=1,\ldots,n,\ k_i=1,\ldots,s_i)$ are convergent series, i.e. holomorphic there.

THEOREM 5.1. *Suppose that one of the conditions is valid:*

(1) *for any number* $i\leq n''$ *with* $p_{ii}>0$ *and for any direction* ℓ *toward* H *in* S(c,r), *there exists a proper domain* c' $(\ell\in c',\ c'\subset c)$ *with respect to all* $\exp((d_{ik_i}-d_{ik_i'})x^{-p_i})$, $k_i,k_i'=1,\ldots,s_i$,

(2) *for any number* $i\leq n''$, $p_i=0$ *or*, $p_i\neq 0$ *and* $d_{ik}\neq d_{ik'}$ *for* $k\neq k'$.

Then, according to the decomposition forms (5.4) *of* A_{i0}'s, $A_i(x)$'s *are* simultaneously decomposable: for any element $\ell \in c$, for some subset c' of c, $\ell \in c'$ and for some $r' \leq r$, there exists an m-by-m invertible matrix P of functions in $\mathcal{A}(c',r')$ such that

$$\lim_{x \to 0} P(x) = I_m ,$$

$$P^{-1}(x)A_i(x)P(x) - x^{p_i}(e_i\partial/\partial x_i)P(x) = \oplus_{k=1}^{s} B_{ik}(x) ,$$

i.e. $$P^{-1}(x)\cdot\nabla_i\cdot P(x) = \oplus_{k=1}^{s}((e_i\partial/\partial x_i) - x^{-p_i}B_{ik}(x)) ,$$

where $B_{ik}(x)$ is an m_k by m_k matrix of functions in $\mathcal{A}(c',r')$ with

$$\lim_{x \to 0} B_{ik} = d_{ik}I_{m_k} + N_{ik} ,$$

or all $i=1,\ldots,n'$.

PROOF. Suppose (1), then, using LEMMA 5.2 repeatedly, we reach the conclusion. Under the condition (2), by the same argument as in the proof and using THEOREMS 4.2 or 4.6, we can arrive at the conclusion. Q.E.D.

REMARK 5.2. If ∇_i and ∇_j are commutative for $i,j \leq n''$, and if one of $d_{i1},\ldots,d_{is},d_{j1},\ldots,d_{js}$ is non-zero, then

(i) $p_{ji} = p_{ij} = 0$, or

(ii) $p_j = p_i$ and $p_{ji}d_{ik} = p_{ij}d_{jk}$ for $k=1,\ldots,s$.

In fact, applying THEOREM 5.1, we obtain, for $k=1,\ldots,s$, differential operators

$$(e_i\partial/\partial x_i) - x^{-p_i}B_{ik}(x) , \quad i=1,\ldots,n' ,$$

commutative each other, i.e.

$$x^{p_i}((e_i\partial/\partial x_i) - p_{ji})B_{jk} + B_{jk}B_{ik} = x^{p_j}((e_j\partial/\partial x_j) - p_{ij})B_{ik} + B_{ik}B_{jk} .$$

For $i,j \leq n''$, by taking the trace of matrices of each side, we see

$$x^{p_i}((e_i\partial/\partial x_i) - p_{ji})(\mathrm{tr}B_{jk}) = x^{p_j}((e_j\partial/\partial x_j) - p_{ij})(\mathrm{tr}B_{ik}) ,$$

from which we deduce that (i) or (ii) is valid.

Moreover,

THEOREM 5.2. *There exists an invertible matrix* $Q(x)$ *such that*

(5.9) $Q(x)$ *is a product of matrices of the forms*

$$I_m+R_1(x_i')x_i+\dots+R_N(x_i')x_i^N \quad \text{and} \quad \oplus_{i=1}^m x_i^{N_i} ,$$

for $i=1,\dots,n''$, *where* $x_i'=(x_1,\dots,x_{i-1},s_{i+1},\dots,x_n)$, $R_1,\dots,R_N$ *are strongly asymptotically developable*, N, $N_1,\dots,N_m$ *are positive integers*,

$$(5.10)\quad Q(x)^{-1}\{x^{-p_i}(B_{ik}(x)Q(x)-(x_i\partial/\partial x_i)Q\} = x^{-q_i}C_i(x)$$

where $q_i=(q_{i1},\dots,q_{in''},0,\dots,0)\in\mathbb{N}^n$, *and* $C_i(x)$ *is strongly asymptotically developable for* $i=1,\dots,n'$, *with the following property*: *if* $q_i=0$ *then* $q_{ji}=0$ *for* $j\neq i$, *and*

$$\lim_{x\to 0}C_i(x) = \oplus_{k_i=1}^{s_i'}(c_{ik_i}I_{m_{k_i}'}+N_{ik_i}') ,$$

with N_{ik_i}'s *are properly upper triangular matrix*, c_{ik}'s *are complex numbers satisfying*

$$c_{ik_i} \neq c_{ik_i'}+\mu \text{ for any integer } \mu \text{ and for any } k_i\neq k_i' .$$

In order prove this theorem, we use the following lemmas:

LEMMA 5.3. *For each* $i=1,\dots,n''$ *with* $p_i=0$ *and for any positive integer* L, *there exists an invertible matrix* $Q_{iL}(x)$ *of functions in* $\mathcal{A}(c,r)$ *of the form*

$$Q_{iL}(x) = \Pi_{h=1}^{L}(I_m+R_{ih}(x_i')x_i^h) ,$$

such that $R_{ih}(x_i')$ *are strongly asymptotically developable with respect to* x_i', *and*

$$Q_{iL}(x)^{-1}\{A_i(x)Q_{iL}(x)-(x_i\partial/\partial x_i)Q_{iL}(x)\}$$

$$= \oplus_{k_i=1}^{s_i}(B_{iLk_ih_ih_i'}(x_i')x_i^{\mu_{ih_ih_i'}})_{h_i,h_i'=1,\dots,t_i}+B_{iL,L}(x)x_i^{L+1} ,$$

where $B_{iLk_ih_ih_i'}(x_i')$ *is an* $m_{k_ih_i}$*-by-*$m_{k_ih_i'}$ *matrix of functions strongly asymptotically developable with respect to* x_i' *for* $h_i, h_i'=1,\dots,t_i$, *in particular* $B_{iLk_ih_ih_i'}(x_i')=0$ *for* $h_i<h_i'$ *and* $B_{iL,L}(x)$ *is an m-by-m matrix of functions in* $\mathcal{A}(c,r)$.

Furthermore, for

$$L=\max\{\mu_{h_ih_i'}: p_i=0,\ h_i,\ h_i'=1,\dots,t_i\}\ ,$$

$$D_i(x_i)^{-1}Q_{iL}(x)^{-1}\{A_i(x)Q_{iL}(x)D_i(x_i)-(x_i\partial/\partial x_i)(Q_{iL}(x)D_i(x_i))\}$$

$$=\oplus_{k_i=1}^{s_i}(B_{i,L,k_ih_ih_i'})_{h_i,h_i'=1,\dots,t_i}+B_{i,L+1}(x)'\ ,$$

where $D_i(x_i)=\oplus_{k_i=1}^{s_i}\oplus_{h_i=1}^{t_i}x_i^{\mu_{h_ih_i}}$, *and* $B_{i,L+1}(x)'$ *is strongly asymptotically developable.*

PROOF OF LEMMA 5.3. We construct such an invertible matrix by an induction on L. Suppose that we obtain the invertible matrix $Q_{i,L}(x)$ and we proceed to construct $Q_{i,L+1}(x)$. First, we rewrite

$$B_{iL,L}(x)=B_{iL,L+1}(x_i')+x_iB_{i,L+1,L+1}(x)\ ,$$

where $B_{i,L,L+1}(x_i')$ is an m-by-m matrix of functions strongly asymptotically developable with respect to x_i' and $B_{i,L+1,L+1}(x)$ is m-by-m matrices of functions in A(c,r). Consider the system of equations with respect to $R_{i,L+1}(x_i')$ and $B_{i,L+1,h_i,h_i'}$, h_i, $h_i'=1,\dots,t_i$ of the form

$$B_{i,L,L+1}(x_i')-B_{i0}(x_i')R_{i,L+1}(x_i')+R_{i,L+1}(x_i')B_{i0}(x_i')+(L+1)R_{i,L+1}(x_i')$$

$$=(B_{i,L+1,k_ik_i'h_ih_i'})_{k_i,k_i'=1,\dots,s_i,h_i,h_i'=1,\dots,t_i}\ ,$$

where

$$B_{i0}(x_i')=\lim_{x_i\to 0}A_i(x)=\lim_{x_i\to 0}Q_{iL}(x)^{-1}A_i(x)Q_{iL}(x)-(x_i\partial/\partial x_i)Q_{iL}(x)\ .$$

Put

$$R_{i,L+1} = (R_{i,L+1,k_ik_i'h_ih_i'})_{k_i,k_i'=1,\ldots,s_i,h_i,h_i'=1,\ldots,t_i}$$

of the blocknized form as $(B_{i,L;1,k_ik_i'h_ih_i'})$. Therefore, by the assumptions on $d_{ik_ih_i}$'s, we can determine $R_{i,L+1,k_ik_i'h_ih_i'}$'s so that

$$B_{i,L+1,k_ik_i'h_ih_i'}=0 \text{ if } d_{ik_ih_i}-d_{ik_i'h_i'}-(L+1)\neq 0.$$

Define $Q_{i,L+1}(x)=Q_{iL}(x)(I_m+R_{i,L+1}(x_i')x_i^{L+1}$. Then, we can verify that this matrix has the property stated above. Q.E.D.

LEMMA 5.4. *Let* $F(x)$ *and* $G(x)$ *be* m-*by*-m *matrices of functions in* $\mathcal{A}(c,r)$ *and let* $p=(p_1,\ldots,p_{n''},0,\ldots,0)\in\mathbb{N}^n$. *Suppose that*

$$(x_i\partial/\partial x_i)(x^{-p}G(x))+x^{-p}G(x)F(x) = (e_j\partial/\partial x_j)F(x)+x^{-p}F(x)G(x) ,$$

and $\lim_{x\to 0}F(x)=dI_m+T$, *where* d *is a complex number and* T *is a properly upper triangular matrix. Then,* $x_i^{-p_i}G(x)$ *is strongly asymptotically developable in* $S(c,r)$.

PROOF OF LEMMA 5.4. As F and G are strongly asymptotically developable, we can write

$$F(x) = F_0(x_i')+x_iF_1(x_i')+\ldots+x_i^{p_i-1}F_{p_i-1}(x_i')+x_i^{p_i}F_{p_i}(x) ,$$

$$G(x) = G_0(x_i')+x_iG_1(x_i')+\ldots+x_i^{p_i-1}G_{p_i-1}(x_i')+x_i^{p_i}G_{p_i}(x) ,$$

where F_0, $G_0,\ldots,F_{p_i-1}$, G_{p_i-1} are strongly asymptotically developable with respect to x_i' and F_{p_i}, G_{p_i} are strongly asymptotically developable with respect to x in $S(c,r)$. Substituting these expansions in the equality, we obtain the equalities

$$(q-p_i)G_q+G_qF_0+\ldots+G_0F_q = F_0G_q+\ldots+F_qG_0 , \quad q=0,\ldots,p_i-1 .$$

Therefore, by induction on q, we see that $G_0=\ldots=G_{p_i-1}=0$. Q.E.D.

PROOF OF THEOREM 5.2. First, apply LEMMA 5.3 to A_i's, and secondly apply LEMMA 5.2 to $Q_{iL}^{-1}\{A_iQ_{iL}-(x_i\partial/\partial x_i)Q_{iL}\}$, and finally apply LEMMA 5.4 to the condition of complete integrability. Q.E.D.

Part III

STOKES PHENOMENA AND RIEMANN-HILBERT-BIRKHOFF PROBLEM

SECTION III.1. INTRODUCTION

Science has been developed to foresee the future on one hand and to know the past on the other, from data which we can observe. The latter is a typical "inverse problem". In mathematics, there are many inverse problems: the Riemann-Hilbert-Birkhoff problem is one of them.

Consider a system of linear homogeneous ordinary differential equations on the Riemann sphere $\mathbb{P}^1$ of the form

$$(S) \qquad \frac{d}{dx} u = A(x)u ,$$

where x is a canonical holomorphic coordinate on $\mathbb{P}^1-\{\infty\}$ and $A(x)$ is an m-by-m matrix of rational functions of x which are holomorphic on the complement of a set $\{a_1,\dots,a_k\}$ of a finite number of points in $\mathbb{P}^1-\{\infty\}$. Denote by H the set $\{a_1,\dots,a_k,a_{k+1}=\infty\}$. If we fix a point $a_0 \in \mathbb{P}^1-H$, then the set V of all germs of local holomorphic solutions at a_0 forms an m-dimensional vector space over $\mathbb{C}$ by the existence theorem of Cauchy. Given any loop ℓ in $\mathbb{P}^1-H$ starting and ending at a_0, the analytic continuation of solutions along ℓ defines an automorphism of V. We can easily see that this automorphism is multiplicative in ℓ and that it depends only on the homotopy class of ℓ in the fundamental group $\pi_1(\mathbb{P}^1-H,\ a_0)$. Thus, we can define a homomorphism

$$\rho:\pi_1(\mathbb{P}^1-H,\ a_0)\longrightarrow GL(m,\mathbb{C}) ,$$

which is the so-called monodromy representation of the system of linear homogeneous differential equations.

Riemann studied the monodromy representation of the hypergeometric equation with three parameters (a,b,c)

$$x(1-x)\frac{d^2}{dx^2}u+\{c-(1+a+b)x\}\frac{d}{dx}u-abx=0\ ,$$

and he proved that all homomorphisms

$$\rho:\pi_1(\mathbb{P}^1-\{0,1,\infty\},\ a_0)\to GL(2,\mathbb{C})$$

can be obtained as the monodromy representation of the hypergeometric equation. Note that this differential equation is <u>regular singular</u> at 0, 1 and ∞: a system (S) is said to be regular singular at a singular point a if (S) has a fundamental solution matrix of solutions of the form $P(x)(x-a)^M$ with $P(x)$ meromorphic at a and M constant. Moreover, he asked whether <u>every finite-dimensional complex representation ρ of the fundamental group $\pi_1(\mathbb{P}^1-H)$ can be obtained as the monodromy representation of a system of differential equations on $\mathbb{P}^1$ with regular singular points on</u> H. In 1900, Hilbert chose this problem as the twenty-first of "Mathematical Problems"; he himself solved it in the case where m=2 and k is arbitrary, and for this reason it was named the Riemann-Hilbert problem. In 1908 Plemelj solved it without restriction.

By studying systems of ordinary differential equations with irregular singular points, G. D. Birkhoff formulated the generalized Riemann-Hilbert problem, which is now called the Riemann-Hilbert-Birkhoff problem [4].

At the singular point a_i, $i=1,\dots,k+1$, we rewrite the equation in the form

$$(x-a_i)^{q_i+1}\frac{d}{dx}u=B_i(x)u$$

(at $a_{k+1}=\infty$, we use the coordinate $t=x^{-1}$). If $q_i>0$ and $B_i(a_i)$ has m distinct eigenvalues, then we have a formal power series transformation $\hat{P}_i(x)$ in $(x-a_i)$ which reduces the system to the canonical system

$$\frac{d}{dx}v_i=\left(\frac{d\Lambda_i(x)}{dx}+\frac{M_i}{x-a_i}\right)v_i$$

where $\Lambda_i(x)$ is an m-by-m diagonal matrix of polynomials of degree (q_i+1) in $(x-a_i)^{-1}$

and M_i is an m-by-m upper triangular matrix commuting with $\Lambda_i(x)$. Moreover, there exists a sectorial covering $\{S_{i,h};\ h=1,\ldots,2q_i\}$ of a punctured disc with the center at a_i such that in each $S_{i,h}$ we have a holomorphic matricial function $P_{i,h}$ asymptotic to $\hat{P}_i(x)$ and that $P_{i,h}(x)(x-a_i)^{M_i}\exp(\Lambda_i(x))$ forms a fundamental solution matrix in $S_{i,h}$. By the property of fundamental solution matrix, there exists a constant matrix $C_{i,hh'}$ such that

$$P_{ih}(x-a_i)^{M_i}\exp(\hat{\Lambda}_i(x))C_{i,hh'} = P_{ih'}(x-a_i)^{M_i}\exp(\Lambda_i(x))$$

in $S_{i,h}\cap S_{i,h'}$. The collection $\{C_{i,hh'}\}$ is called the Stokes multipliers at a_i. Conversely, let there be given consistently an m-dimensional representation ρ of $\pi_1(\mathbb{P}^1-H)$ and

$$\{q_i,\Lambda_i,M_i,C_{i,hh'};\ i=1,\ldots,k+1,\ h,h'=1,\ldots,2q_i\}\ .$$

Birkhoff asked whether we can construct a system (S) which has the given data. He solved this problem in 1913 [5]. He constructed a local equation at each given singular point and patched them together to obtain a global equation.

We now go on to consider systems of linear homogeneous ordinary differential equations with coefficients of meromorphic functions on a Riemann surface M with poles at most on a set H of a finite number of points on M. In this case, we have no global coordinate x at our disposal, so we express a global system of differential equations as a Pfaffian system

$$du - \Omega u = 0\ ,$$

where Ω is an m-by-m matrix of global meromorphic 1-forms on M with poles at most on H. We can consider the Riemann-Hilbert problem and the Riemann-Hilbert-Birkhoff problem on M also.

The Riemann-Hilbert problem asks whether the following is a surjective mapping:

$$\left\{\begin{array}{l}\text{systems of Pfaffian equations with} \\ \text{regular singular points on } H\end{array}\right\} \longrightarrow \{\text{representations of } \pi_1(M-H)\} .$$

By using the theory of vector bundles, in 1956 Röhrl [68] solved the problem on any Riemann surface M (admitting apparent singular points). Röhrl constructed his solution by three steps of procedures. (This idea is essentially the same as Birkhoff's.)

(1) The set of all conjugate classes of m-dimensional complex representations of $\pi_1(M-H)$ is isomorphic to the set of isomorphism classes of flat vector bundles of rank m over M-H, i.e., the first cohomology set with coefficients in the constant sheaf $GL(m,\mathbb{C})$ over M-H:

$$\mathrm{Hom}(\pi_1(M-H),\ GL(m,\mathbb{C}))/\widehat{\mathrm{conj}}. \simeq H^1(\ M-H,\ GL(m,\mathbb{C}))\ .$$

(Notice that the right-hand side is also regarded as the set of isomorphism classes of locally constant sheaves of rank m over M-H.)

(2) For a flat vector bundle $\mathfrak{E}$ together with the natural connection d on M-H, there exists a holomorphic vector bundle $\mathfrak{F}$ over M together with an integrable connection ∇ with regular singular points on H such that

a) the restriction on M-H of $\mathfrak{F}$ is isomorphic to $\mathfrak{E}$: denote by i the isomorphism,

b) the following diagram is commutative:

$$\begin{array}{ccc} & \mathfrak{F}|_{M-H} & \xrightarrow{\ i\ } & \mathfrak{E} \\ & \nabla|_{M-H}\downarrow & \circlearrowright & \downarrow d \\ \mathfrak{F}\otimes\Omega^1_M|_{M-H} = & \mathfrak{F}|_{M-H}\otimes\Omega^1_{M-H} & \xrightarrow{\ i\otimes 1\ } & \mathfrak{E}\otimes\Omega^1_{M-H} \end{array}$$

where Ω^1_M and Ω^1_{M-H} are the sheaves of germs of holomorphic 1-forms on M and M-H, respectively.

Roughly speaking, these imply the following: let $\{U_\alpha\}$ be an "adequate" open covering of M-H. Let $\{C_{\alpha\beta}\}$ and $\{P_{\alpha\beta}\}$ be the collections of transition functions of $\mathcal{G}$ and the restriction of $\mathcal{F}$ with respect to the covering, respectively. Let $\{\Omega_\alpha\}$ be the collection of connection matrices of the restriction of the connection ∇ on M-H with respect to the covering, so that

$$dC_{\alpha\beta} = 0, \; dP_{\alpha\beta} = \Omega_\alpha P_{\alpha\beta} - P_{\alpha\beta}\Omega_\beta \quad \text{in} \quad U_\alpha \cap U_\beta \;.$$

Then there exists a collection $\{\Phi_\alpha\}$ of invertible m-by-m matricial functions with respect to the covering $\{U_\alpha\}$ of M-H such that

$$\Phi_\alpha C_{\alpha\beta} = P_{\alpha\beta}\Phi_\beta \;, \quad d\Phi_\alpha = \Omega_\alpha\Phi_\alpha - \Phi_\alpha \cdot 0 \;.$$

(3) The holomorphic vector bundle $\mathcal{F}$ admits m independent meromorphic global sections, so $\mathcal{F}$ is "meromorphically" trivial and the connection ∇ coincides with a "homomorphism" defined by a global Pfaffian system $(d-\Omega)u=0$. Broadly speaking, there exists a collection $\{Q_\alpha\}$ of m-by-m invertible meromorphic matricial functions with respect to the covering such that

$$Q_\alpha P_{\alpha\beta} = Q_\beta \quad \text{in} \quad U_\alpha \cap U_\beta \;.$$

By the equalities, $Q_\alpha\Phi_\alpha C_{\alpha\beta} = Q_\beta\Phi_\beta$, and so we can define a meromorphic matricial function Ω on M-H by

$$\Omega = d(Q_\alpha\Phi_\alpha)(Q_\alpha\Phi_\alpha)^{-1} \quad \text{on} \quad U_\alpha \text{ for all } \alpha.$$

Then, $$dQ_\alpha = \Omega Q_\alpha - Q_\alpha\Omega_\alpha \quad \text{on} \quad U_\alpha \text{ for all } \alpha,$$

and Ω can be prolongated meromorphically to the whole M (for a precise construction, see Section III.4).

We can verify (1) by a persevering computation. The second is verified by the existence of local systems of differential equations which have the local monodromy representation. The key to the proof of (3) is the Kodaira-Nakano vanishing theorem in the case where M is compact and is the Oka-Grauert theorem in the case where M is noncompact, i.e., open (hence a Stein manifold).

The fact (1) is gradually understood in a more general context. From the perspective of algebraic geometry, in 1969 Deligne defined the regular singularity along a divisor H for the several-variables case and solved the Riemann-Hilbert problem in the sense of existence of connection, i.e., corresponding to (2). He constructed his connection by using Hironaka's resolution theorem and Manin's local construction of connection with regular singularities on a normal crossing divisor. In the one-variable case, the divisor is a discrete set, so the procedure (2) is a completely local one. But in the several-variables case, where we can slide along H from one point to another, we have to patch together local connections. Deligne proved, moreover, that there exists a natural one-to-one correspondence between each pair of the following:

(A) {Isomorphism classes of locally free sheaves of $\mathcal{O}(*H)$-modules of rank m together with an integrable connection regular singular on H over M}

(B) {conjugate classes of m-dimensional representation of $\pi_1(M-H)$}

(C) $H^1(M-H,GL(m,\mathbb{C}))$

(D) {isomorphism classes of flat vector bundles over M-H}

(E) {isomorphism classes of locally constant sheaves over M-H}

(F) {isomorphism classes of locally free sheaves of $\mathcal{O}_{M-H}$-modules of rank M together with an integrable connection over M-H}.

Several authors solved the Riemann-Hilbert problem in the several-variables case as existence of global completely integrable Pfaffian systems of which the monodromy representation coincides with the given representation of $\pi_1(M-H)$: Gérard [13], Suzuki [77], Kita [37], etc.

After Birkhoff, there was little in the literature concerning the Riemann-Hilbert-Birkhoff problem except for Trjitzinsky [80] until the 1970's. By deeply studying linear ordinary differential equations of the second order with polynomial coefficients, Y. Sibuya [72] proved that, for a given family of invertible matrices with special properties, there exist second order linear ordinary differential equations of which the Stokes multipliers at $\infty \in \mathbb{P}^1$ coincide with the given family of matrices (by using Nevanlinna's theorem). In a series of papers [33, 2, 3], Jurkat-

Lutz et al. studied explicitly complete formal and analytic invariants of systems of linear homogeneous ordinary differential equations at an irregular singular point "modulo transformations", in other words, a classification of systems of linear homogeneous ordinary differential equations. In order to assert that their analytic invariants are complete, we have to solve the Riemann-Hilbert-Birkhoff problem of local version: given a system of analytic invariants, to construct a system of linear homogeneous ordinary differential equations with an irregular singularity at the given point and whose analytic invariants coincide with the given ones.

Suggested by the work of Birkhoff [5] and Cartan (cf. [19]), Y. Sibuya provided a powerful tool for the Riemann-Hilbert-Birkhoff problem of local version and he himself formulated a classification of systems of linear homogeneous ordinary differential equations. The tool is a kind of vanishing theorem as follows. Let $\{S(c_h,r)\}$ be a sectorial open covering of a punctured disc $D(r)-\{0\}$ at the origin in $\mathbb{C}$, where the c_h's are open sets in $[0,2\pi)\simeq S^1=\{|z|=1\}$. Given a family $\{P_{hh'}\}$ of m-by-m invertible matricial functions such that

i) $P_{hh'}$ is holomorphic and asymptotic to the unit matrix I_m in $S(c_h\cap c_{h'},r)$,

ii) $P_{hh'}P_{h'h''} = P_{hh''}$ in $S(c_h\cap c_{h'}\cap c_{h''},r)$,

then there exists a family $\{Q_h\}$ of m-by-m invertible matricial functions such that

iii) Q_h is holomorphic and asymptotically developable in $S(c_h,r')$

iv) $Q_hP_{hh'} = Q_{h'}$ in $S(c_h\cap c_{h'},r')$

for some $r'\leq r$ (see also Section I.1)

Let $\Lambda(x)$ be an m-by-m matrix of polynomials in x and let M by an m-by-m constant matrix such that $d\Lambda/dx$, Λ and M are commutative with each other. Take a sufficiently fine open covering $\{c_h\}$ of S^1. Denote by $\mathcal{S}(\{c_h\},\Lambda,M)$ the set of all family $\{C_{hh'}\}$ of constant m-by-m matrices such that

(a) $C_{hh'}C_{h'h''} = C_{hh''}$

(b) $\exp(\Lambda(x))x^M C_{hh'}[\exp(\Lambda(x))x^M]^{-1}$ is holomorphic and asymptotically developable to I_m in $S(c_h \cap c_{h'}, r)$ for some r.

Denote by $\hat{E}(\Lambda,M)$ the set of all systems of differential equations which is formally equivalent to the system

$$\frac{d}{dx} u = \left(\frac{d\Lambda}{dx} + \frac{M}{x} \right) u .$$

<u>Theorem</u> A. $\hat{E}(\Lambda,M)/\sim \;\simeq \mathcal{S}(\{c_h\},\Lambda,M)$,

where for two systems (S_1) $\frac{d}{dx} u = A(x)u$ and (S_2) $\frac{d}{dx} v = B(x)v$, $(S_1) \sim (S_2)$ if they are holomorphically equivalent, i.e., there exists a holomorphic transformation $u=P(x)v$ which changes (S_1) into (S_2) [74]. In the above study, Sibuya introduced the presheaf of germs of functions asymptotically developable [73].

Stimulated by works of Jurkat-Letz et al. and that of Sibuya, B. Malgrange definitively introduced the sheaf of germs of functions asymptotically developable and formulated another classification of germs of systems of linear homogeneous ordinary differential equations with an irregular singular point at the origin. For an open set c in S^1 and a positive number r, denote by $G(\Lambda,M)(c,r)$ the set of all matricial functions P such that

(c) P is holomorphic and asymptotic to I_m in $S(c,r)$,

(d) P is a solution to the following system there:

$$(*) \qquad \frac{d}{dx} P = \left(\frac{d\Lambda}{dx} + \frac{M}{x} \right) P - P \left(\frac{d\Lambda}{dx} + \frac{M}{x} \right) .$$

Then, together with the natural restrictions, $\{G(\Lambda,M)(c,r)\}$ is an inductive system. Put

$$G(\Lambda,M)(c) = \operatorname{dir.lim}_{r \to 0} G(\Lambda,M)(c,r) .$$

Thus, we have a presheaf $\{G(\Lambda,M)(c), i_{cc'}\}$, where $i_{cc'}$ is the natural restriction mapping. Denote by $G(\Lambda,M)$ the associated sheaf over S^1. Then,

<u>Theorem</u> B. $\hat{E}(\Lambda,M)/\sim \simeq H^1(S^1, G(\Lambda,M))$.

This formulation is essentially the same as Sibuya's, because the solution P(x) to the system (*) is of the form

$$P(x) = \exp(\Lambda(x))x^M C\{\exp(\Lambda(x))x^M\}^{-1} ,$$

where C is a constant matrix.

The purpose of this Part is to generalize these results not only in the local sense but also in the global sense. We formulate the Riemann-Hilbert-Birkhoff problem in the several-variables case based on the results of Part II, and we solve it by using the vanishing theorem stated in Part I.

SECTION III.2. FUNDAMENTAL MATRICES OF SOLUTIONS OF COMPLETELY INTEGRABLE LINEAR PFAFFIAN SYSTEMS.

Consider a completely integrable system of Pfaffian equations with singularities in $H=\{x\in\mathbb{C}^n:\ x_1\ldots x_{n''}=0\}$: $(d-\Omega)u=0$, where

$$\Omega=\sum_{i=1}^{n''}x^{-p_i}x_i^{-1}A_i(x)dx_i+\sum_{i=n''+1}^{n}x^{-p_i}A_i(x)dx_i\ ,$$

satisfying $d\Omega=\Omega\wedge\Omega$, and $p_i=(p_{i1},\ldots,p_{in''},0,\ldots,0)\in\mathbb{N}^n$, $A_i(x)$ is an m-by-m matrix of functions holomorphic and strongly asymptotically developable in an open polysector $S(c,r)=\Pi_{i=1}^{n''}S(c_i,r_i)\times\Pi_{i=n''+1}^{n}D(r_i)$. Namely, consider the system of differential equations of the first order $\nabla_i u=0$, $i=1,\ldots,n$, where

$$\nabla_i=(e_i\partial/\partial x_i)-x^{-p_i}A_i(x)\ ,\quad i=1,\ldots,n,$$

satisfying $\nabla_i\nabla_j=\nabla_j\nabla_i$, $i,j=1,\ldots,n$, here, $e_i=x_i$ $(i\leq n'')$, $e_i=1$ $(i>n'')$.

First, we consider the case that $m=1$.

PROPOSITION 2.1. *The solution of the system is of the form:* $cw(x)x^a\exp(x^{-p}h(x))$, *where* c *is a constant*, $a=(a_1,\ldots,a_{n''},0,\ldots,0)\in\mathbb{C}^n$, $p=(p_1,\ldots,p_{n''},0,\ldots,0)\in\mathbb{N}^n$, $w(x)$ *and* $h(x)$ *are holomorphic and strongly asymptotically developable in* $S(c,r)$.

PROOF. The condition $d\Omega=\Omega\wedge\Omega$ implies that

$$(e_i\partial/\partial x_i)(x^{-p_j}A_j(x))=(e_j\partial/\partial x_j)(x^{-p_i}A_i(x))\ ,\quad i,j=1,\ldots,n.$$

Put $u(x)=x^a\exp(x^{-p}g(x))$. Then, a and $x^{-p}g(x)$ satisfy the following completely integrable system of differential equations

$$(e_i\partial/\partial x_i)(x^{-p}g(x))=x^{-p_i}A_i(x)-a_i\ ,\quad i=1,\ldots,n,$$

i.e. $((e_i\partial/\partial x_i)-p_i)g(x) = x^{p-p_i}A_i(x)-a_ix^p$, $i=1,\dots,n$.

Define $P=(\max\{P_{i1}: i=1,\dots,n''\}, \dots,\max\{p_{in''}: i=1,\dots,n''\},0,\dots,0)$, and define a_i as the coefficient $TA(x^{p-p_i}A_i)_p$ of strongly asymptotic expansion of $x^{p-p_i}A_i$, $i=1,\dots,n$. Then, by the completely integrability, we can easily see that this system has a formal series solution of x. Using the THEOREM II.4.1 repeatedly, we can prove that this system has a unique family G of formal solutions in S(c,r) and a unique solution g with TA(g)=G in S(c,r). Put $h(x)=App_p(g;x)$ and $k(x)=x^{-P}(g(x)-h(x))$, then h and k are strongly asymptotically developable. Finally, we write $w(x)=\exp(k(x))$ and we conclude that $cw(x)x^a\exp(h(x))$ is a solution of $(d-\Omega)u=0$ with a constant c.
Q.E.D.

REMARK 2.1. If $A_i(x)$'s are holomorphic in a polydisc $D(r)=\Pi_{i=1}^{n}D(r_i)$, then w(x) and h(x) are holomorphic there.

In the case that $m>1$, we restrict ourselves here to give a theorem under one of the following conditions:

(1) for some $i=1,\dots,n''$, $p_i=0$ and $A_{i0}=\lim_{x\to 0}A_i(x)$ has m distinct eigenvalues and each difference of two eigenvalues are not a integer,

(2) for some $i=1,\dots,n''$, P_{ii} is positive, A_{i0} has m distinct eigenvalues d_{ik}'s $(k=1,\dots,m)$ and for any direction ℓ in c toward H there exists a polysector which contains ℓ and which is non-negatively proper domain with respect to all functions $\exp((d_{ik}-d_{ik'})x^{-p_i})$, $k,k'=1,\dots,m$,

(3) for all $i=1,\dots,n''$, $p_i=0$ or, $p_{ii}>0$ and A_{i0} has m distinct eigenvalues,

THEOREM 2.1. Suppose one of the conditions (1), (2) and (3). Then, for any direction ℓ toward H in S(c,r), there exists a proper open subpolysector S(c',r) containing ℓ such that in S(c',r) the fundamental matrix of solutions of $(d-\Omega)u=0$ is of the form $P(x)x^T\exp(x^{-p}H(x))$, where P(x) is an invertible m-by-m matrix of holomorphic and strongly asymptotically developable in S(c',r), T is an n''-ple of upper triangular matrices, $x^T=\Pi_{i=1}^{n''}x_i^{T_i}$, $p=(p_1,\dots,p_{n''},0\dots,0)\in\mathbb{N}^n$ and H(x) is a di-

agonal matrix of holomorphic and strongly asymptotically developable in $S(c',r)$, *and* T_i's *and* $H(x)$ *are commutative each other*.

PROOF. By using THEOREMS II.5.2, 1 and PROPOSITION II.5.1, we can deduce this theorem. Q.E.D.

REMARK 2.2. If $p_i=0$ for all $i=1,\ldots,n''$, the fundamental matrix of solution is reduced to be $P(x)x^T$ and $P(x)$ can be taken as matricial functions holomorphic and strongly asymptotically developable in the full polysector $S(c,r)$: the key is the uniqueness theorem (THEOREM II.4.1) of system of equations with simple singularities.

REMARK 2.3. By REMARK II.5.2, $x^{-p}H(x)$ can be taken of the form $\sum_{k=1}^{t} x^{-q_k} H_k(x)$, where $H_k(x)$'s $(k=1,\ldots,t)$ are like $H(x)$ and $q_k=(q_{k1},\ldots,q_{kn''},0,\ldots,0)$ $\in \mathbb{N}^n$ $(k=1,\ldots,t)$ such that at most one of $q_{1i},\ldots,q_{ki}$ is non-zero for all $i=1,\ldots,n''$.

REMARK 2.4. In case (3), if, in addition, any difference of two eigenvalues of A_{i0} is not a integer for i with $p_i=0$, then we can prove Theorem 2.1 without using Theorem II.5.2 and so, $P(x)$ belongs to $GL(m, \mathcal{A}'^{-}(S(c',r')))$. In other case, in general, $P(x)$ belongs to $GL(m, \mathcal{A}^{-}(*H)(S(c',r')))$.

SECTION III.3. STOKES PHENOMENA AND RIEMANN-HILBERT-BIRKHOFF PROBLEM OF LOCAL VERSION.

Consider a Pfaffian system $(d-\Omega)u=0$ in a polydisc $D(r)=\Pi_{i=1}^{n}D(r_i)$ at the origin in $\mathbb{C}^n$ with singular points in $H=\{x\in\mathbb{C}^n: x_1\ldots x_{n''}=0\}$, where

$$\Omega = \sum_{i=1}^{n''} x^{-p_i} x_i^{-1} A_i(x)dx_i + \sum_{i=n''+1}^{n} x^{-p_i} A_i(x)dx_i ,$$

$p_i=(p_{i1},\ldots,p_{in''},0,\ldots,0)\in\mathbb{N}^n$, $A_i(x)$ is an m-by-m matrix of holomorphic functions in $D(r)$. Suppose that for one of i's, $p_{ii}>0$ and $A_i(0)$ has m distinct eigenvalues. Then, applying THEOREM 2.1 to this system we can assert that there exists upper triangular matrices T_i's $(i=1,\ldots,n'')$, $p=(p_1,\ldots,p_{n''},0,\ldots,0)\in\mathbb{N}^n$ and a diagonal matrix $H(x)$ of holomorphic functions such that T_i's and $H(x)$ are commutative each other, and for any direction ℓ toward H, there exists a polysector S and an m-by-m invertible matrix $P(x)$ holomorphic and strongly asymptotically developable in S, and $P(x)\Pi_{i=1}^{n''}x_i^{T_i}\exp(x^{-p}H(x))$ forms a fundamental matrix of solutions for the system $(d-\Omega)u=o$ in S. And so, by the property of fundamental matrix of solutions, we see

PROPOSITION 3.1. *There exists a sectorial covering*

$$\{S_k(r) = \Pi_{i=1}^{n''} S(c_{ik_i},r_i)\times\Pi_{i=n''+1}^{n}D(r_i);\ k=(k_1,\ldots,k_{n''})\le K\in\mathbb{N}^{n''}\}$$

of $D(r)-H$ *such that*

(1) *there exists an invertible* m-by-m *matrix* $P_k(x)$ *holomorphic and strongly asymptotically developable in* $S_k(r)$ *for all* k,

(2) *there exists a family* $\{C_{kk'}: k,k'\le K\}$ *of constant invertible* m-by-m *matrices satisfying*

$$P_k(x)\Pi_{i=1}^{n''}x_i^{T_i}\exp(x^{-p}H(x))C_{kk'} = P_{k'}(x)\Pi_{i=1}^{n''}x_i^{T_i}\exp(x^{-p}H(x))$$

in $S_k(r)\cap S_{k'}(r)\neq\emptyset$, $k,k'\le K$.

We traditionally call the family $\{C_{kk'}\}$ the Stokes' multipliers and we see easily that $\{C_{kk'}\}$ satisfies the cocycle condition:

$$C_{kk'}C_{k'k''}C_{k''k} = I_m \text{ in } S_k(r)\cap S_{k'}(r)\cap S_{k''}(r) \neq \emptyset ,$$

namely $\{C_{kk'}\}$ is 1-cocycle of the covering $\{S_k(r)\}$ with coefficients in the sheaf $GL(m,\mathbb{C})$. Denote by $[\{C_{kk'}\}]\in H^1(\{S_k(r)\},GL(m,\mathbb{C}))$ the class of cohomology of $\{C_{kk'}\}$. Then, this class is uniquely determined for the given Pfaffian system.

Suppose now that, for any $i=1,\ldots,n''$, $p_i=0$ or, $p_{ii}>0$ and $A_i(0)$ has m distinct eigenvalues. Then, under the above notation, $FA(P_k)$ belongs to $\mathcal{O}_{D(r)}\hat{|}_H(D(r))$ and $FA(P_k)=FA(P_{k'})$ for $k,k'\leq K$.

Denote by $D(r)^-$ the real blow up along H of D(r) and denote by pr the projection. By the construction, there exists a covering $\{S_k(r)^-: k\leq K\}$ of $D(r)^-$ such that $pr(S_k(r)^-)-H=S_k(r)$ for all $k\leq K$. Set, for $k,k'\leq K$,

$$F_{kk'} = P_k^{-1}P_{k'} = (x^T\exp(x^{-p}H(x)))C_{kk'}(x^T\exp(x^{-p}H(X))^{-1} ,$$

then $\{F_{kk'}\}$ is a 1-cocycle of the covering $\{S_k(r)^-\}$ with coefficients in $GL(m,\mathcal{A})_{I_m}$, and the class of cohomology $[\{F_{kk'}\}]\in H^1(\{S_k(r)^-\},GL(m,\mathcal{A})_{I_m})$ is uniquely determined. Note that $F_{kk'}$ satisfies the system of Pfaffian equations $dF_{kk'}-\Omega_c F_{kk'}+F_{kk'}\Omega_c=0$, where $\Omega_c=d(x^{-p}H(x))+\sum_{i=1}^{n''}T_i x_i^{-1}dx_i$. Conversely, if $F_{kk'}$ is a solution of the system $dF_{kk'}-\Omega_c F_{kk'}+F_{kk'}\Omega_c=0$, then $F_{kk'}$ is written of the form $F_{kk'}=E_c C_{kk'}E_c^{-1}$ for some constant matrix $C_{kk'}$, where E_c is a fundamental matrix of solutions of the system of Pfaffian equations $(d-\Omega_c)u=0$, e.g. $E_c=x^T\exp(x^{-p}H(x))$. Denote by $\mathcal{K}er\mathcal{H}om(d-\Omega_c)I_m$ the subsheaf of $GL(m,\mathcal{A}^-)_{I_m}$ over $D(r)^-$ of germs of invertible matricial functions F satisfying the system $dF-\Omega_c F+F\Omega_c=0$. Then, $\{F_{kk'}\}$ is a 1-cocycle of the covering $\{S_k(r)^-\}$ with coefficients in $\mathcal{K}er\mathcal{H}om(d-\Omega_c)$.

Conversely, let be given

(G1) a matrix $\Omega_c=d(x^{-p}H(x))+\sum_{i=1}^{n''}T_i x_i^{-1}dx_i$, where T_i is an m-by-m constant matrix for $i=1,\ldots,n''$ and $x^{-p}H(x)$ is an m-by-m matrix of meromorphic functions with poles at most on H, such that T_i's and $x^{-p}H(x)$ are commutative each other,

(G2) a 1-cocycle $\{F_{kk'}\}$ of the covering $\{S_k(r)^-\}$ with coefficients in the sheaf $\mathcal{K}er\mathcal{H}om(d-\Omega_c)$.

PROPOSITION 3.3. *For the given* (G1) *and* (G2), *there exists a completely integrable Pfaffian system* $(d-\Omega)u=0$ *on* $D(r')$ *for* $r' \leq r$ *such that*

(i) Ω *is an* m-*by*-m *matrix of meromorphic* 1-*forms on* $D(r')$ *with* poles at most on H,

(ii) *for any* $k \leq K$, *there exists an* m-*by*-m *invertible matrix* $P_k(x)$ *of functions in* $\mathcal{A}(S_k(r))$ *with which* $P_k(x)x^T\exp(x^{-p}H(x))$ *forms a fundamental matrix of solutions of the system* $(d-\Omega)u=0$, i.e.

$$dP_k = \Omega P_k - P_k\Omega_c \ .$$

PROOF. By THEOREM I.3.4, there exists 0-cochain

$$\{P_k:\ P_k \in GL(m, \mathcal{A}(S_k)),\ k \leq K\},$$

of the covering $\{S_k(r')^-\}$ with coefficients in the restricted sheaf of $\mathcal{K}er\mathcal{H}om(d-\Omega_c)$ on $D(r')^-$ such that

$$P_k^{-1}P_{k'} = F_{kk'} = (x^T\exp(x^{-p}H(x))C_{kk'}(x^T\exp(x^{-p}H(x))^{-1} \ ,$$

in $S_k(r')\cap S_{k'}(r')\neq\emptyset$. Put $Q_k(x) = P_k(x)x^T\exp(x^{-p}H(x))$ for $x \in S_k(r')$, then

$$dQ_k(x)Q_k(x)^{-1} = dQ_{k'}(x)Q_{k'}(x)^{-1}$$

for $x \in S_k(r')\cap S_{k'}(r')\neq\emptyset$. Define a matrix Ω of holomorphic 1-forms in $D(r)-H$ by putting $\Omega(x)=dQ_k(x)Q_k(x)^{-1}$ for $x \in S_k(r')$. Then, by the construction, we see easily that Ω is meromorphic in $D(r')$. i.e. at most with poles on H. Q.E.D.

Let T_i (i=1,...,n") and $x^{-p}H(x)$ be as above. Denote by $CONN(T_1,\ldots,T_{n''},x^{-p}H(x))$ the set of all connections on the stalk of $\mathcal{O}(*H)^m$ at 0, i.e. $\mathbb{C}$-linear mappings satisfying Leibnitz' rule

$$\nabla:\ (\mathcal{O}(*H)^m)_0 \longrightarrow (\Omega^1(*H)^m)_0 \ ,$$

such that the extension of ∇ to $(\mathcal{O}_{D(r)}\hat{|}_H(*H)^m)_0$ coincides with the connection ∇_c defined by $(d-\Omega_c)$, in other words, there exists a free basis $<e_1',\ldots,e_m'>$ for $(\mathcal{O}_{D(r)}\hat{|}_H(*H)^m)_0$ such that

$$\nabla<e_1',\ldots,e_m'> = <e_1',\ldots,e_m'>(d-\Omega_c) .$$

As we see it above, there exists a canonical mapping μ from $\mathrm{CONN}(T_1,\ldots,T_{n''},x^{-P}H(x))$ into $H^1(pr^{-1}(0), \mathcal{K}er\mathcal{H}om(d-\Omega_c)_{I_m}|_{pr^{-1}(0)})$ and μ is a surjective mapping.

THEOREM 3.1. *The mapping* μ *is a bijection.*

PROOF. We shall prove μ is injective. Let $\{F_{kk'}\}$ and $\{G_{kk'}\}$ be two 1-cocycle of the covering $\{c_k\}$ of $pr^{-1}(0)$ with coefficients n the sheaf $\mathcal{K}er\mathcal{H}om(d-\Omega_c)_{I_M}|_{pr^{-1}(0)}$. As we see in the proof of Proposition 3.2, there exist a 0-cocycle $\{P_k\}$ and $\{R_k\}$ of $\{c_k\}$ with coefficients in $\mathcal{K}er\mathcal{H}om(d-\Omega_c)_{I_m}|_{pr^{-1}(0)}$ such that $F_{kk'}=P_k^{-1}P_{k'}$ and $G_{kk'}=R_k^{-1}R_{k'}$ in $c_k\cap c_{k'}\neq\emptyset$, respectively. Denote by Ω and Ω' the germs of 1-form defined by $\Omega=dP_kP_k^{-1}+P_k\Omega_cP_k^{-1}$ and $\Omega'=dR_kR_k^{-1}+R_k\Omega_cR_k^{-1}$ on c_k, respectively. Suppose that there exists 0-cochain $\{z_k\}$ of $\{c_k\}$ with coefficients in $\mathcal{K}er\mathcal{H}om(d-\Omega_c)_{I_m}|_{pr^{-1}(0)}$ such that $F_{kk'}=Z_kG_{kk'}Z_{k'}^{-1}$ in $c_k\cap c_{k'}\neq\emptyset$. Then, we see easily that

$$P_kZ_kR_k^{-1} = P_{k'}Z_{k'}R_{k'}^{-1} \text{ in } c_k\cap c_{k'}\neq\emptyset.$$

Hence, we can define a function Z on $pr^{-1}(0)$ by putting $Z=P_kZ_kR_k^{-1}$ on c_k, and Z is thought to be an element in $GL(m,\mathcal{O}(*H)_0)$. By the definition, $dZ=\Omega Z-Z\Omega'$. This implies that $(d-\Omega)$ and $(d-\Omega')$ define the same connection on $(\mathcal{O}(*H)^m)_0$. Q.E.D.

SECTION III.4. STOKES PHENOMENA AND RIEMANN-HILBERT-BIRKHOFF PROBLEM OF GLOBAL VERSION.

In this section, we keep the notation used in Section I.3.

Let $\mathcal{S}$ be a locally free sheaf of $\mathcal{O}(*H)$-modules of rank m and let ∇ be an integrable connection on $\mathcal{S}$. For any point $p \in H$, there exists an open set U in M containing p and a free basis $e_U=(e_{1U},\ldots,e_{mU})$ of $\mathcal{S}$ over U. With respect to the free basis e_U, the connection ∇ is represented by $(d+\Omega_{eU})$, i.e.

$$\nabla(\langle e_{1U},\ldots,e_{mU}\rangle u) = \langle e_{1U},\ldots,e_{mU}\rangle(du + \Omega_{eU}u) ,$$

where Ω_{eU} is an m-by-m matrix of meromorphic 1-forms with poles at most on H and u is any m-vector of functions in $\mathcal{O}(*H)(U)$. If $f_U=\langle f_{1U},\ldots,f_{mU}\rangle$ is another free basis of $\mathcal{S}$ over U, then there exists an m-by-m invertible matrix G of functions in $\mathcal{O}(*H)(U)$ such that

$$\langle f_{1U},\ldots,f_{mU}\rangle = \langle e_{1U},\ldots,e_{mU}\rangle G ,$$

$$\nabla(\langle f_{1U},\ldots,f_{mU}\rangle u) = \langle f_{1U},\ldots,f_{mU}\rangle(du + (G^{-1}\{\Omega_{eU}G+dG\})u) .$$

Let $x_1,\ldots,x_m$ be holomorphic local coordinates at p on U with $U\cap H=\{x_1\ldots x_{n''}=0\}$, then Ω_{eU} is written of the form

$$\Omega_{eU} = \sum_{i=1}^{n''} x^{-p_i} x_i^{-1} A_i(x)dx_i + \sum_{i=n''+1}^{n} x^{-p_i} A_i(x)dx_i ,$$

where $p_i=(p_{i1},\ldots,p_{in''},0\ldots,0)\in \mathbb{N}^n$ and $A_i(x)$ is an m-by-m matrix of holomorphic functions in U for $i=1,\ldots,n$, and Ω_{eU} satisfies, by the integrability condition, $d\Omega_{eU}+\Omega_{eU}\wedge\Omega_{eU}=0$.

Suppose that for any point p on H

(H#) there exists an open set U containing p with holomorphic coordinates $x_1,\ldots,x_n$ and a free basis $\langle e_{1U},\ldots,e_{mU}\rangle$ of $\mathcal{S}$ such that Ω_{eU} is written of the above form satisfying

(H#1) $p_i=0$ or, $p_i>0$ and $A_i(0)$ has m distinct eigenvalues for all $i=1,\dots,n''$.

Then, by Theorem 2.1 and the same consideration as in Section III.2, we can assert

Theorem 4.1. If the assumption (H#) is satisfied for any point p on H, then

(DATA1) for any point p on H and for an open set U containing p, there exists an m-by-m matrix $\Omega_{p,c}$ of meromorphic 1-forms with poles at most on $H\cap U$, which is written of the form

$$\Omega_{p,c} = dD(x(p)) + \sum_{i=1}^{n''} T_i(p)x_i(p)dx_i(p) \; ,$$

where $x_1(p),\dots,x_n(p)$ are holomorphic local coordinates on U at p with $U\cap H=\bigcup_{i=1}^{n''}\{x_i(p)=0\}$, $D(x(p))$ is an m-by-m diagonal matrix of functions in $\mathcal{O}(*H)(U)$ and $T_i(p)$, $i=1,\dots,n''$, are upper triangular matrices such that $D(x(p))$, $T_i(p)$ $(i=1,\dots,n'')$ are commutative each other,

(DATA2) there exists a locally free sheaf $\mathcal{F}$ of $\mathcal{A}^-(*H)$-modules over M^- a connection $\nabla_{\mathcal{F}}$ on $\mathcal{F}$ such that

(#1) there exists an isomorphism $g:\mathcal{F}\longrightarrow pr^*\mathcal{S}\otimes_{pr^*\mathcal{O}(*H)}\mathcal{A}^-(*H)$ such that $g^{-1}\cdot(\nabla\otimes id)\cdot g=\nabla_{\mathcal{F}}$,

(#2) for any point p on H, there exists an open set U containing p such that the isomorphism class $[\mathcal{F}|_{U^-}]$ of $\mathcal{F}$ restricted on $U^-=pr^{-1}(U)$ belongs to $H^1(U^-,\mathcal{K}er\,\mathcal{H}om(d-\Omega_{p,c})_{I_m}))$.

In other words, we can assert

PROPOSITION 4.1. If the assumption (H#) is satisfied for any point p on H, then there exists a locally free sheaf $\mathcal{F}$ of $\mathcal{A}^-(*H)$-modules over M^- and a connection $\nabla_{\mathcal{F}}$ on $\mathcal{F}$ such that

(i) there exists an isomorphism $g:\mathcal{F}\longrightarrow pr^*\mathcal{S}\otimes_{pr^*\mathcal{O}(*H)}\mathcal{A}^-(*H)$ such that

$g^{-1}\cdot(\nabla\otimes id)\cdot g=\nabla_{\mathcal{F}}$,

(ii) for any point p on H, there exists an open set U containing p such that the isomorphism class $[\mathcal{F}|_{U^-}]$ of $\mathcal{F}$ restricted on $U^-=pr^{-1}(U)$ belongs to the subset $H^1(U^-,GL(m,\mathcal{A}^-)_{I_m}|_{U^-})$ of $H^1(U^-,GL(m,\mathcal{A}^-(*H)|_{U^-})$.

(iii) for any point p on H and for an open set U containing p with holomorphic coordinates $x_1(p),\ldots,x_n(p)$, $U\cap H=\bigcup_{i=1}^{n''}\{x_i(p)=0\}$, there exist an m-by-m diagonal matrix $D(x(p))$ of functions in $\mathcal{O}(*H)(U)$ and upper triangular matrices $T_i(p)$, $i=1,\ldots,n''$ such that

(iii.a) $D(x(p))$, $T_i(p)$ $(i=1,\ldots,n'')$ are cummutative each other,

(iii.b) for any point p' in $pr^{-1}(p)$ there exists an open set $V^-(p')$ containing p' and a free basis $\langle e(V^-(p'))_1,\ldots,e(V^-(p'))_m\rangle$ for $\mathcal{F}$ on $V^-(p')$ such that

$$\nabla_{\mathcal{F}}(\langle e(V^-(p'))_1,\ldots,e(V^-(p'))_m\rangle v) = (\langle e(V^-(p'))_1,\ldots,e(V^-(p'))_m\rangle(d+\Omega_{p,c})v)$$

where v is any m-vector of functions in $\mathcal{A}^-(*H)(V^-)$.

Conversely, we can prove the following.

Theorem 4.2. For any locally free sheaf $\mathcal{F}$ of $\mathcal{A}^-(*H)$-modules over M^- and an integrable connection $\nabla_{\mathcal{F}}$ on $\mathcal{F}$ satisfying (ii) and (iii), there exists a locally free sheaf $\mathcal{S}$ of $\mathcal{O}(*H)$-modules over M and an integrable connection ∇ on $\mathcal{S}$ satisfyig (i).

In order to prove Theorem 4.2, it suffices to prove the following lemma.

Lemma 4.1. For a locally free sheaf $\mathcal{F}$ of $\mathcal{A}^-(*H)$-modules over M^- and an integrable connection $\nabla_{\mathcal{F}}$ on $\mathcal{F}$ satisfying (ii), there exists a locally free sheaf $\mathcal{S}$ of $\mathcal{O}(*H)$-modules over M and an integrable connection ∇ on $\mathcal{S}$ satisfying (i).

PROOF OF LEMMA 4.1. Let $\{V^-(p'):p'\in pr^{-1}(p),\ p\in M\}$ be an open covering of M^-, where $V^-(p')$ is an open set containing p'. For any p', take a basis $\langle e(V^-(p'))_1,\ldots,e(V^-(p'))_m\rangle$ for $\mathcal{F}$ on $V^-(p')$. Then, we obtain a 1-cocycle

$\{F_{V^-(p')V^-(q')}\}$ of the covering with coefficients in $GL(m, \mathcal{A}^-(*H))$, which forms a family of transition functions of $\mathcal{F}$ relative to the covering and a family of connection matrices $\{\Omega(V^-(p'))\}$ for $\nabla_{\mathcal{F}}$ relative to the covering and the bases. Suppose that each $V^-(p')$ is sufficiently small. Then by (ii) and Theorem I.3.4, for any $p \in H$, there exists a 0-cochain $\{p_{V^-(p')}\}$ of the covering $\{V^-(p'): p' \in pr^{-1}(p)\}$ with coefficients in $GL(m, \mathcal{A}^-)\big|_{U^-(p)}$, where $U^-(p) = \bigcup_{p' \in pr^{-1}(p)} V^-(p')$, such that $F_{V^-(p')V^-(q')} = P_{V^-(p')}{}^{-1} P_{V^-(q')}$ on $V^-(p') \cap V^-(q') \neq \emptyset$. Put $P_{V^-(p')} = 0$ for p', $pr(p') \notin H$. Then, by the cocycle condition, we see that

$$P_{V^-(p')} F_{V^-(p')V^-(q')} P_{V^-(q')}{}^{-1} = P_{V^-(p'')} F_{V^-(p'')V^-(q'')} P_{V^-(q'')}{}^{-1},$$

for $p = pr(p') = pr(p'')$ and $q = pr(q') = pr(q'')$. And so, we can define a function $G_{U^-(p)U^-(q)}$ on $U^-(p) \cap U^-(q)$ by putting

$$G_{U^-(p)U^-(q)} = P_{V^-(p')} F_{V^-(p')V^-(q')} P_{V^-(q')}{}^{-1} \text{ on } V^-(p') \cap V^-(q') .$$

Moreover, $G_{U^-(p)U^-(q)}$ can be thought as a meromorphic function $G_{U(p)U(q)}$ on $U(p) \cap U(q)$, where $U(b) = pr(U^-(b))$ for $b = p, q$. Thus, we obatin a 1-cocycle $\{G_{U(p)U(q)}\}$ of the covering $\{U(p)\}$ of M with coefficients in $GL(m, \mathcal{O}(*H))$. This 1-cocycle defines a locally free sheaf $\mathcal{S}$ of $\mathcal{O}(*H)$-modules over M. Moreover, we see that

$$\{dP_{V^-(p')} + P_{V^-(p')} \Omega_{V^-(p')}\} P_{V^-(p')}{}^{-1} = \{dp_{V^-(p'')} + P_{V^-(p'')} \Omega_{V^-(p'')}\} P_{V^-(p'')}{}^{-1},$$

on $V^-(p') \cap V^-(p'')$ for $p = pr(p') = pr(p'')$. And so, we can define a 1-form $\Omega_{U^-(p)}$ on $U^-(p)$ by putting

$$\Omega_{U^-(p)} = \{dP_{V^-(p')} + P_{V^-(p')} \Omega_{p,c}\} P_{V^-(p')}{}^{-1}$$

which is thought as a meromorphic 1-form $\Omega_{U(p)}$ on $U(p)$. By the definitions we can verify that on $U(p) \cap U(q)$

$$dG_{U(p)U(q)} = \Omega_{U(p)} G_{U(p)U(q)} - G_{U(p)U(q)} \Omega_{U(q)} .$$

Hence, the family $\{\Omega_{U(p)}\}$ defined an integrable connection ∇ on the locally free sheaf $\mathcal{S}$. Q.E.D.

Consider now a locally free sheaf $\mathcal{S}$ of $\mathcal{O}$-modules and an integrable connection ∇ on $\mathcal{S} \otimes_{\mathcal{O}} \mathcal{O}(*H)$. Suppose that for any point p on H,

(H#') there exists an open set U containing p with holomorphic coordinates $x_1, \ldots, x_n$ and a free basis $\langle e_{1U}, \ldots, e_{mU} \rangle$ of S such that Ω_{eU} is written of the above form satisfying

(H#2) $p_i = 0$ and any difference of two eigenvalues of $A_i(0)$ is not a integer or, $p_i > 0$ and $A_i(0)$ has m distinct eigenvalues for all $i = 1, \ldots, n$".

In this case, by Remark 1.4 and by the same consideration as above, we can assert

<u>Theorem 4.3</u>. <u>If the assumption</u> (H#') <u>is satisfied for any point on</u> p, <u>then</u> $\mathcal{F}$ <u>can be taken as a locally free sheaf of</u> $\mathcal{A}^-$<u>-modules in Theorem 4.1</u>.

<u>Proposition 4.2</u>. <u>If the assumption</u> (H#') <u>is satisfied for any point</u> p <u>on</u> H, <u>there exists a locally free sheaf</u> $\mathcal{F}$ <u>of</u> $\mathcal{A}^-$<u>-modules and</u> an integrable connection on $\mathcal{F} \otimes_{\mathcal{A}^-} \mathcal{A}^-(*H)$ satisfying (ii), (iii) and

(i') <u>there exists an isomorphism</u> $g: \mathcal{F} \longrightarrow pr^* \mathcal{S} \otimes_{pr^* \mathcal{O}} \mathcal{A}^-$ <u>such that</u> $(g \otimes_{\mathcal{A}^-} id) \cdot (\nabla \otimes_{pr^* \mathcal{O}} id) \cdot (g \otimes_{\mathcal{A}^-} id)^{-1} = \nabla_{\mathcal{F}}$.

In the same way as in the proof of Lemma 4.1, we can prove

<u>Lemma 4.2</u>. <u>For a locally free sheaf</u> $\mathcal{F}$ <u>of</u> $\mathcal{A}^-$<u>-modules over</u> M^- <u>and an integrable connection</u> $\nabla_{\mathcal{F}}$ <u>on</u> $\mathcal{F} \otimes_{\mathcal{A}^-} \mathcal{A}^-(*H)$ <u>satisfying</u> (ii), <u>there exists a locally free sheaf</u> $\mathcal{S}$ <u>of</u> $\mathcal{O}$<u>-modules over</u> M <u>and an integrable connection</u> ∇ <u>on</u> $\mathcal{S} \otimes_{\mathcal{O}} \mathcal{O}(*H)$ <u>satisfying</u> (i').

And so, in particular, we obtain

Theorem 4.4. *For any locally free sheaf $\mathcal{F}$ of $\mathcal{A}^-$-modules over M^- and an integrable connection $\nabla_{\mathcal{F}}$ on $\mathcal{F}\otimes_{\mathcal{A}^-}\mathcal{A}^-(*H)$ satisfying (i) and (iii), there exists a locally free sheaf $\mathcal{S}$ of $\mathcal{O}$-modules over M and an integrable connection ∇ on $\mathcal{S}\otimes_{\mathcal{O}}\mathcal{O}(*H)$ satisfying (i').*

REMARK 4.1. In the preceding argument, we can replace $\mathcal{A}^-$ by $\mathcal{A}'^-$.

For the above $\mathcal{F}$ and $\nabla_{\mathcal{F}}$, the kernel sheaf $\mathcal{K}er\,\nabla_{\mathcal{F}}$ is a local constant sheaf $\mathcal{C}$ on M^- which is thought to be a locally constant sheaf on $M-H$. For any $p\in H$ and for any $p'\in pr^{-1}(p)$, take an open set $V^-(p')$ as in (iii.b), then $\{V^-(p'): p'\in pr^{-1}(p), p\in H\}$ is an open covering of a neighborhood of $pr^{-1}(H)$. Then, by (iii),

$$c(V^-(p')) = \langle e(V^-(p')_1,\dots,e(V^-(p')_m\rangle ESS(x(p))$$

is a free basis for $\mathcal{C}$ over $V^-(p')$, where $x(p)$ is the holomorphic local coordinates system chosen at $p=pr(p')$ and $ESS(x(p))$ is a fundamental matrix of solutions of the system of equations

$$du+\{dD(x(p))+\textstyle\sum_{i=1}^{n''}T_i(p)dx_i(p)\}u = 0 ,$$

say, $ESS(x(p))=\exp(D(x(p)))\prod_{i=1}^{n''}x_i(p)^{T_i(p)}$. For $p', q'\in pr^{-1}(H)$, denote by $C_{V^-(p')V^-(q')}$ the transition matrix for $\mathcal{C}$ relative to the bases $c(V^-(p'))$, $c(V^-(q'))$, i.e.

$$c(V^-(p'))C_{V^-(p')V^-(q')} = c(V^-(q')) .$$

And so, the matricial function

$$G_{V^-(p')V^-(q')} = ESS(x(p))C_{V^-(p')V^-(q')}ESS(x(q))^{-1}$$

is the transition function for $\mathcal{F}$ relative to $e(V^-(p'))$, $e(V^-(q'))$. Therefore $G_{V^-(p')V^-(q')}$ is strongly asymptotically developable in $pr(V^-(p')\cap V^-(q'))-H$. In particular, if $pr(p')=pr(q')$, $G_{V^-(p')V^-(q')}$ is strongly asymptotically developable to I_m. Conversely, given a locally free sheaf $\mathcal{C}$ over M-H and the matricial function ESS(x(p)) for any $p\in H$ satisfying the above properties, there exists a locally free sheaf $\mathcal{F}$ over M^- of $\mathcal{A}'^-$-modules and an integrable connection $\nabla_{\mathcal{F}}$ on $\mathcal{F}\otimes_{\mathcal{A}'^-}\mathcal{A}'^-(*H)$ that (iii.b) is satisfied and such that the kernel sheaf $\mathcal{K}er\,\nabla_{\mathcal{F}}$ coincides with the given locally constant sheaf $\mathcal{C}$. And so, by Theorem 4.4, there exists a locally free sheaf $\mathcal{S}$ over M of $\mathcal{O}$-modules and an integrable connection ∇ on $\mathcal{S}\otimes_{\mathcal{O}}\mathcal{O}(*H)$ satisfying (i) for this $(\mathcal{F},\nabla_{\mathcal{F}})$ constructed from $\mathcal{C}$ and ESS(x(p)) for any $p\in H$. Moreover, if M is a Stein manifold or a projective manifold, by using Oka-Cartan's Theorem or Kodaira's vanishing Theorem, we can prove the following (cf. [68], [14], [77]).

<u>Theorem 4.5</u>. <u>There exists a divisor</u> H' <u>on</u> M <u>and an integrable connection</u> ∇ <u>on the sheaf</u> $\mathcal{O}(*(H+H'))^m$, i.e. <u>a completely integrable system of Pfaffian equations on</u> M <u>with irregular singular points on</u> (H+H'), <u>such that</u> (i) <u>is satisfied</u>.

PROOF. Let $\{G_{U(p)U(q)}\}$ be the 1-cocycle of the covering $\{U(p)\}$ with coefficients in $GL(m,\mathcal{O})$, which defines the locally free sheaf S stated in Theorem 4.4. It remains to prove that for some divisor H' there exists a 0-cochain $\{R_{U(p)}\}$ of the covering $\{U(p)\}$ with coefficients in $GL(m,\mathcal{O}(*(H'))$ such that $G_{U(p)U(q)}=R_{U(p)}^{-1}R_{U(q)}$ on $U(p)\cap U(q)$. For this purpose, it suffices to prove that there exist m independent global <u>meromorphic</u> sections on the holomorphic vector bundle defined by the 1-cocycle $\{G_{U(p)U(q)}\}$. This can be proven by the usual techniques, i.e. complex blowing up and vanishing theorem (cf. [21], [62], [68], [77]). Q.E.D.

This theorem is classically formulated and proven by G. D. Birkhoff [4], [5] and reformulated locally by Balser-Jurkat-Lutz [2], [3], Sibuya [73], [74] and Malgrange [56] in one variable case.

Part IV

∇-POINCARÉ'S LEMMA AND ∇-DERHAM COHOMOLOGY THEOREM

SECTION IV. 1. INTRODUCTION

In the field of differential geometry, de Rham established a beautiful isomorphism theorem of cohomology, the de Rham cohomology theorem (cf. [21]). Namely, let X be a real C^∞-manifold of dimension n and let $\mathcal{E}^q$ be the sheaf of germs of C^∞-q-forms on X for q=0,1,...,n. Then, for any q=0,1,...,n, we have the isomorphism

Theorem A. $H^q(\Gamma(X,\mathcal{E}^\cdot),d) \simeq H^q(X,\mathbb{C})$,

where d is the exterior derivative and the left-hand side is the q-th cohomology group of the complex

$$\Gamma(X, \mathcal{E}^0) \xrightarrow{d} \Gamma(X, \mathcal{E}^1) \xrightarrow{d} \dots \xrightarrow{d} \Gamma(X, \mathcal{E}^n) \longrightarrow 0$$

consisting of the groups $\Gamma(X, \mathcal{E}^\cdot)$ of global sections of $\mathcal{E}^\cdot$ on X together with the exterior derivative d. As is well known, one of the keys to the proof is Poincaré's lemma: the following sequence of sheaves is exact

$$\mathcal{E}^0 \xrightarrow{d} \mathcal{E}^1 \xrightarrow{d} \dots \xrightarrow{d} \mathcal{E}^n \longrightarrow 0 .$$

Another key fact is that $\mathcal{E}^q$, q=0,1,...,n are fine sheaves.

In the field of complex analysis or in the field of algebraic geometry, analogues of the de Rham cohomology theorem have been established by several authors, for example, Serre (cf. [19, 21]), Grothendieck [22] and Deligne [11]. The purpose of this Part is to provide a new analogue of the de Rham cohomology theorem. Before stating our problem, however, we explain the results of the preceding authors.

Let M be a complex analytic manifold and let $H=\sum H_h$ be a divisor on M. Denote by $\mathcal{O}=\Omega^0,\Omega^1,\dots,\Omega^n$ the sheaves over M of germs of holomorphic functions, holomorphic 1-forms, ..., holomorphic n-forms, respectively, and denote by $\mathcal{O}(*H)=$

$\Omega^0(*H), \Omega^1(*H), \ldots, \Omega^n(*H)$ the sheaves over M of germs of meromorphic functions meromorphic 1-forms, ..., meromorphic n-forms with poles at most on H, respectively. Let $\mathcal{S}$ be a locally free sheaf over M of $\mathcal{O}(*H)$-modules of rank m and let ∇ be an integrable connection on $\mathcal{S}$. We will view ∇ as a homomorphism of abelian sheaves

$$\nabla: \mathcal{S} \longrightarrow \mathcal{S} \otimes_{\mathcal{O}(*H)} \Omega^1(*H)$$

which satisfies Leibniz' rule

$$\nabla(f \cdot u) = u \otimes df + f \wedge \nabla u$$

for any local sections $f \in \mathcal{O}(*H)(U)$, $u \in \mathcal{S}(U)$ over any open set U in M, and which extends to define a structure of complex on $\mathcal{S} \otimes_{\mathcal{O}(*H)} \Omega^{\cdot}(*H)$, the "$\nabla$-de Rham complex" of $(\mathcal{S}, \nabla)$

$$\mathcal{S} \xrightarrow{\nabla} \mathcal{S} \otimes_{\mathcal{O}(*H)} \Omega^1(*H) \xrightarrow{\nabla} \cdots \xrightarrow{\nabla} \mathcal{S} \otimes_{\mathcal{O}(*H)} \Omega^n(*H) \longrightarrow 0 .$$

For any point $p \in M$, there exist an open set U in M containing p and a free basis $e_U = \langle e_{1U}, \ldots, e_{mU} \rangle$ of $\mathcal{S}$ over U. With respect to the free basis e_U, the connection ∇ is represented by $(d + \Omega_{e_U})$, i.e.,

$$\nabla(\langle e_{1U}, \ldots, e_{mU} \rangle u) = \langle e_{1U}, \ldots, e_{mU} \rangle (du + \Omega_{e_U} u) ,$$

where Ω_{e_U} is an m-by-m matrix of meromorphic 1-forms with poles at most on H and u is any m-vector of functions in $\mathcal{O}(*H)(U)$. We call the matrix Ω_{e_U} the <u>connection matrix</u> with respect to e_U over U of ∇. If $f_U = \langle f_{1U}, \ldots, f_{mU} \rangle$ is another free basis of $\mathcal{S}$ over U, then there exists an m-by-m invertible matrix G of functions in $\mathcal{O}(*H)(U)$ such that

$$\langle f_{1U}, \ldots, f_{mU} \rangle = \langle e_{1U}, \ldots, e_{mU} \rangle G ,$$

$$\nabla(\langle f_{1U}, \ldots, f_{mU} \rangle u) = \langle f_{1U}, \ldots, f_{mU} \rangle (du + G^{-1}\{\Omega_{e_U} G + dG\} u) .$$

Let $x_1, \ldots, x_n$ be holomorphic local coordinates at p on U, then Ω_{e_U} is written in the form

$$\Omega_{e_U} = \sum_{i=1}^{n} T_i(x)dx_i ,$$

where $T_i(x)$ is an m-by-m matrix of meromorphic functions in U at most with poles on $U \cap H$ for $j=1,\ldots,n$ and Ω_{e_U} satisfies, by the integrability conditions,

$$d\Omega_{e_U} + \Omega_{e_U} \wedge \Omega_{e_U} = 0 .$$

In the case in which M is a Stein manifold, $H=\emptyset$, $\mathcal{S}=\mathcal{O}$ and $\nabla=d$, J.-P. Serre proved the so-called "holomorphic de Rham cohomology theorem".

<u>Theorem</u> B. For $q=0,1,\ldots,n$,

$$H^q(\Gamma(M,\Omega^\cdot),d) \simeq \mathbf{H}^q(M,(\Omega^\cdot,d)) \simeq H^q(M,\mathbb{C}) ,$$

where the left hand side is the q-th cohomology group of the complex

$$\Gamma(M,\mathcal{O}) \xrightarrow{d} \Gamma(M,\Omega^1) \xrightarrow{d} \cdots \xrightarrow{d} \Gamma(M,\Omega^n) \longrightarrow 0$$

consisting of the abelian groups of global sections of $\Omega^\cdot$ on M and the middle member is the q-th hypercohomology group of the de Rham complex for $(\mathcal{O},d)$. The key to the proof that the left-hand side is isomorphic to the middle member is the "vanishing theorem"

$$H^q(M,\Omega^p) = 0 \qquad q \geq 1,\ p \geq 0$$

due to Oka-Cartan (cf. [19, 21, 25, 62]). The key to the proof that the right-hand side is isomorphic to the middle member is the so-called "holomorphic Poincaré's lemma": the following sequence of sheaves is exact

$$\mathcal{O} \xrightarrow{d} \Omega^1 \xrightarrow{d} \cdots \xrightarrow{d} \Omega^n \longrightarrow 0 .$$

Notice that the constant sheaf $\mathbb{C}$ can be regarded as the kernel sheaf of $d:\mathcal{O} \rightarrow \Omega^1$. In the same way (cf. Deligne [11]), we can assert

$$H^q(\Gamma(M,\mathcal{S}\otimes\Omega^\cdot),\nabla) \simeq H^q(M,\mathbb{C}^m) ,$$

for $q=0,1,\ldots,n$, if $H=\emptyset$ in the above set-up. Here, we use the fact that the kernel sheaf $\mathcal{K}er\,\nabla$ of $\nabla:\mathcal{S}\to\mathcal{S}\otimes_{\mathcal{O}}\Omega^1$ is isomorphic to the constant sheaf $\mathbb{C}^m$, because of the Frobenius theorem.

Now we pass to the results of Grothendieck and Deligne. We restrict ourselves here to explaining the analytic results of their work, from which, by using the GAGA principle and Hironaka's resolution theorem, we deduce precise algebraic theorems [22, 11, 21].

Suppose that M is a compact complex analytic manifold, that $H=\sum H_h$ is a positive divisor at most with normal crossings, and that the complement M-H of H is a Stein manifold. Denote by i the natural inclusion mapping from M-H to M. The sheaf $i_*i^*\Omega^q$ is regarded as the sheaf of germs of q-forms holomorphic in M-H and essentially singular on H for $q=0,1,\ldots,n$. We write $\Omega^{\cdot}_{M-H}$ for the restricted sheaves of $\Omega^{\cdot}$ on M-H and d_{M-H} for the restricted exterior differential of d on M-H.

Grothendieck treated the case in which $\mathcal{S}=\mathcal{O}(*H)$ and $\nabla=d$ [21,22].

<u>Theorem</u> C. <u>For any</u> $q=0,1,\ldots,n$, <u>the following are isomorphic to each other:</u>

(a) $H^q(\Gamma(M,\Omega^{\cdot}(*H)),d)$

(b) $\mathbb{H}^q(M,(\Omega^{\cdot}(*H),d))$

(c) $\mathbb{H}^q(M,(i_*i^*\Omega^{\cdot},d))$

(d) $\mathbb{H}^q(M-H,(\Omega^{\cdot}_{M-H},d_{M-H}))$

(e) $H^q(\Gamma(M-H,\Omega^{\cdot}_{M-H}),d_{M-H})$

(f) $H^q(M-H,\mathcal{K}er\,d)=H^q(M-H,\mathbb{C}^1)$.

By Serre's theorem, (d), (e) and (f) are isomorphic to each other, and (c) is trivially isomorphic to (d). The key to the proof that (a) is isomorphic to (b) is the fact that

$$H^q(M,\Omega^p(*H))=0 \quad \text{for} \quad q\geq 1,\ p\geq 0\ ,$$

which is a corollary of Kodaira's vanishing theorem. The key to the proof that (b) is isomorphic to (c) is a kind of "Poincaré's lemma", i.e., that the complex

$(\Omega^{\cdot}(*H),d)$ of sheaves is quasi-isomorphic to $(i_* i^* \Omega^{\cdot},d)$ by the natural inclusion, which is equivalent to the fact that the complex of quotient sheaves $(i_* i^* \Omega^{\cdot}/\Omega^{\cdot}(*H),d)$ is acyclic, i.e., the derived cohomology sheaves are zero.

Let M and H be the same as Grothendieck's cases. Deligne [11] considers the de Rham complex $(\mathcal{S} \otimes_{\mathcal{O}(*H)} \Omega^{\cdot}(*H), \nabla)$ in the case where ∇ is regular singular along H, i.e., for any point $p \in H$, there exist a neighborhood U of p with holomorphic coordinates $x_1,\ldots,x_n$, $U \cap H = \{x_1,\ldots,x_{n''}=0\}$ and a free basis e_U for $\mathcal{S}$ over U such that the connection matrix Ω_{e_U} is of the form of logarithmic poles

$$\Omega_{e_U} = \sum_{i=1}^{n''} x_i^{-1} A_i(x)dx_i + \sum_{i=n''+1}^{n} A_i(x)dx_i ,$$

where the $A_i(x)$'s are m-by-m matrices of holomorphic functions in U. Deligne proves a theorem similar to Grothendieck's. In fact, if we replace $\Omega^{\cdot}$, $\Omega^{\cdot}(*H)$, $i_* i^* \Omega^{\cdot}$, $\mathbb{C}^1$ and d by $\mathcal{S} \otimes_{\mathcal{O}} \Omega^{\cdot}$, $\mathcal{S} \otimes_{\mathcal{O}(*H)} \Omega^{\cdot}(*H)$, $\mathcal{S} \otimes_{\mathcal{O}(*H)} i_* i^* \Omega^{\cdot}$, $\mathbb{C}^m$ and ∇, respectively, in Theorem C, we obtain the theorem of Deligne [11].

Now we explain our problem. Let ∇ be an integrable connection with irregular singular points. We propose to study the cohomology proups of complex

$$\Gamma(M,\mathcal{S}) \xrightarrow{\nabla} \Gamma(M, \mathcal{S} \otimes_{\mathcal{O}(*H)} \Omega^1(*H)) \xrightarrow{\nabla} \ldots \xrightarrow{\nabla} \Gamma(M, \mathcal{S} \otimes_{\mathcal{O}(*H)} \Omega^n(*H)) \to 0$$

consisting of abelian groups of global sections of $\mathcal{S} \otimes_{\mathcal{O}(*H)} \Omega^{\cdot}(*H)$ on M together with the integrable connection ∇ with irregular singular points. According to Katz' report [35], this is Monsky's problem. Suppose as above that H is a normal crossing divisor. Let M^- be the real blow-up of M along H together with the natural projection $pr: M^- \to M$, and let $\mathcal{A}_0^-$ be the sheaf over M^- of germs of functions strongly asymptotically developable to 0 (see Part I). Denote by ∇_0^- the following natural integrable connection induced from

$$\nabla_0^-: \mathcal{A}_0^- \otimes_{pr^*\mathcal{O}} pr^* \mathcal{S} \longrightarrow \mathcal{A}_0^- \otimes_{pr^*\mathcal{O}} pr^*(\mathcal{S} \otimes_{\mathcal{O}(*H)} \Omega^1(*H)) .$$

Denote by $\mathcal{K}er\, \nabla_0^-$ the kernel sheaf of ∇_0^-. Assume that for any point $p \in H$, there exist a neighborhood U of p with holomorphic coordinates $x_1,\ldots,x_n$, $U \cap H = \{x_1 \ldots x_{n''}=0\}$

and a free basis e_U for $\mathcal{S}$ over U such that the connection matrix is of the form

$$\Omega_{e_U} = \sum_{i=1}^{n''} x_1^{-p_{i1}} \cdots x_{n''}^{-p_{in''}} x_i^{-1} A_i(x) dx_i + \sum_{i=n''+1}^{n} x_1^{-p_{i1}} \cdots x_{n''}^{-p_{in''}} A_i(x) dx_i$$

with the following property: for all $i=1,\ldots,n''$,

$$(*) \quad \begin{cases} (1) & p_{i1}=\cdots=p_{in''}=0 \text{ and } A_i(0) \text{ has no eigenvalue of integer} \\ \text{or} & \\ (2) & p_{ii}>0 \text{ and } A_i(0) \text{ is invertible.} \end{cases}$$

Then we can prove

<u>Theorem</u> D. $H^1(M^-, \mathcal{K}er\,\nabla_0^-) \simeq H^1(\Gamma(M, \mathcal{S}\otimes_{\mathcal{O}(*H)}\Omega^{\cdot}(*H)),\nabla)$.
One of the keys to the proof is the vanishing theorem for the commutative case stated in Part I. Another is the ∇-Poincaré's lemma for the complex of sheaves

$$\mathcal{A}^-(*H)\otimes_{pr^*\mathcal{O}(*H)} pr^*\mathcal{S} \xrightarrow{\nabla^-} \mathcal{A}^-(*H)\otimes_{pr^*\mathcal{O}(*H)} pr^*(\mathcal{S}\otimes_{\mathcal{O}(*H)}\Omega^1(*H)) \xrightarrow{\nabla^-} \cdots ,$$

proven by using the results of Part II. For the higher cohomology, we have not yet proven the same result, because the higher cohomologies $H^q(M^-, \mathcal{A}_0^-)$ $(q\geq 2)$ have in general complicated structures. Nevertheless, we conjecture that the isomorphism theorem is valid for $q\geq 2$. This is a problem to be solved in the future. (See [86].)

B. Malgrange [56] proved a theorem of duality. Suppose $M=\mathbb{P}^1$, $H=\{a_1,\ldots,a_k\}\subset M$. Let $\mathcal{S}^*$ be the dual of $\mathcal{S}$ and let ∇^* be the dual connection of ∇, i.e.,

$$\nabla^*: \mathcal{S}^* \longrightarrow \mathcal{S}^*\otimes_{\mathcal{O}(*H)}\Omega^1(*H)$$

$$d\langle f,g\rangle = \langle \nabla f,g\rangle + \langle f,\nabla^* g\rangle$$

$f\in\mathcal{S}(U)$, $g\in\mathcal{S}^*(U)$ for any open set U in M. Denote by ∇_0^{*-} the dual connection of ∇_0^-

$$\nabla_0^{*-}: \mathcal{A}_0^-\otimes_{pr^*\mathcal{O}} pr^*\mathcal{S}^* \longrightarrow \mathcal{A}_0^-\otimes_{pr^*\mathcal{O}}(\mathcal{S}^*\otimes_{\mathcal{O}(*H)}\Omega^1(*H))$$

and denote by $\mathcal{K}er\,\nabla_0^{*-}$ the kernel sheaf of ∇_0^{*-}. Then, without any restriction, we have

<u>Theorem</u> E. <u>For</u> $q=0,1,2$, <u>the dual of</u> $H^q(\Gamma(M,\Omega^\cdot(*H)),\nabla)$ <u>is naturally isomorphic to</u> $H^{2-q}(M^-, \mathcal{K}er\nabla_0^{*-})$.

The author was led to Theorem D by substituting $q=1$ in Theorem E. Note that, in our proof of Theorem D, condition (*) is essentially utilized even in the case of one variable. In the proof of Theorem E, this condition is not assumed.

SECTION IV.2. ASYMPTOTIC ∇-POINCARÉ'S LEMMA.

In this section, we keep the notation used in Section I.3 and I.4.

Let $\mathcal{S}$ be a locally free sheaf of $\mathcal{O}$-modules or $\mathcal{O}(*H)$-modules of rank m on M. Define the locally free sheaf $\mathcal{S}\Omega^q(*H)$ of $\mathcal{O}(*H)$-modules of rank m over M by

$$\mathcal{S}\Omega^q(*H) = \mathcal{S}\otimes_{\mathcal{O}}\Omega^q(*H) ,$$

for q=0,...,n. For q=0, instead of $\mathcal{S}\Omega^0(*H)$, we use frequently $\mathcal{S}(*H)$.

Let ∇ be a connection on $\mathcal{S}(*H)$: ∇ is an additive mapping

$$\nabla: \mathcal{S}(*H) \longrightarrow \mathcal{S}(*H)\otimes_{\mathcal{O}(*H)}\Omega^1(*H) = \mathcal{S}(*H)\otimes_{\mathcal{O}}\Omega^1 = \mathcal{S}\otimes_{\mathcal{O}}\Omega^1(*H) = \mathcal{S}\Omega^1(*H)$$

satisfying Leibniz' rule

$$\nabla(f\cdot u) = u\otimes df + f\wedge\nabla(u)$$

for all sections $f\in\mathcal{O}(*H)(U)$, $u\in\mathcal{S}\Omega^1(*H)(U)$. We suppose that the connection is integrable, that is, the composite mapping

$$\nabla^2: \mathcal{S}(*H) \longrightarrow \mathcal{S}\Omega^1(*H) \longrightarrow \mathcal{S}\Omega^2(*H)$$

is a zero mapping.

Define the locally sheaves $\mathcal{S}^-\Omega^q(*H)$, $\mathcal{S}\mathcal{C}\mathcal{F}^-\Omega^q(*H)$, $\mathcal{S}'^-\Omega^q(*H)$, $\mathcal{S}_{\hat{H}}^-\Omega^q(*H)$ and $\mathcal{S}_0^-\Omega^q$ over the real blow-up M^- of M along the normal crossing divisor H with the natural projection $pr:M^-\longrightarrow M$, by the followings

$$\mathcal{S}^-\Omega^q(*H) = \mathcal{A}^-\otimes_{pr^*\mathcal{O}}pr^*(\mathcal{S}\Omega^q(*H)) ,$$

$$\mathcal{S}\mathcal{C}\mathcal{F}^-\Omega^q(*H) = \mathcal{C}\mathcal{F}^-\otimes_{pr^*\mathcal{O}}pr^*(\mathcal{S}\Omega^q(*H)) ,$$

$$\mathcal{S}'^-\Omega^q(*H) = \mathcal{A}'^-\otimes_{pr^*\mathcal{O}}pr^*(\mathcal{S}\Omega^q(*H)) ,$$

$$\mathcal{S}_{\hat{H}}^-\Omega^q(*H) = pr^*(\mathcal{O}_{M|\hat{H}}\otimes_{\mathcal{O}}\mathcal{S}\Omega^q(*H)) ,$$

and $$\mathcal{S}_0^-\Omega^q = \mathcal{A}_0^-\otimes_{pr^*\mathcal{O}}pr^*(\mathcal{S}\Omega^q(*H)) = \mathcal{A}_0^-\otimes_{pr^*\mathcal{O}}pr^*(\mathcal{S}\otimes_{\mathcal{O}}\Omega^q) ,$$

respectively, for q=0,1,...,n. For q=0, we use frequently $\mathcal{S}^-(*H)$, $\mathcal{S}e\mathcal{F}^-(*H)$, $\mathcal{S}'^-(*H)$, $\mathcal{S}_{\hat{H}}^-(*H)$ and $\mathcal{S}_0^-$ instead of them. Then, by a natural way, we obtain integral connections

$$(2.1)\quad \nabla^- : \mathcal{S}^-(*H) \longrightarrow \mathcal{S}^-\Omega^1(*H) ,$$

$$(2.2)\quad \nabla^- : \mathcal{S}e\mathcal{F}^-(*H) \longrightarrow \mathcal{S}e\mathcal{F}^-\Omega^1(*H) ,$$

$$(2.3)\quad \nabla'^-: \mathcal{S}'^-(*H) \longrightarrow \mathcal{S}'^-\Omega^1(*H) ,$$

$$(2.4)\quad \nabla'^-: \mathcal{S}_{\hat{H}}^-(*H) \longrightarrow \mathcal{S}_{\hat{H}}^-\Omega^1(*H) ,$$

and
$$(2.5)\quad \nabla_0^- : \mathcal{S}_0^- \longrightarrow \mathcal{S}_0^-\Omega^1 .$$

For simplicity, we use also ∇ instead of ∇^-, ∇'^-, ∇_0^-. By the integrability, we can consider the complexes of sheaves

$$(2.6)\quad \mathcal{S}^-(*H) \xrightarrow{\nabla} \mathcal{S}^-\Omega^1(*H) \xrightarrow{\nabla} \cdots \xrightarrow{\nabla} \mathcal{S}^-\Omega^n(*H) \longrightarrow 0$$

$$(2.7)\quad \mathcal{S}e\mathcal{F}^-(*H) \xrightarrow{\nabla} \mathcal{S}e\mathcal{F}^-\Omega^1(*H) \xrightarrow{\nabla} \cdots \xrightarrow{\nabla} \mathcal{S}e\mathcal{F}^-\Omega^n(*H) \longrightarrow 0$$

$$(2.8)\quad \mathcal{S}'^-(*H) \xrightarrow{\nabla} \mathcal{S}'^-\Omega^1(*H) \xrightarrow{\nabla} \cdots \xrightarrow{\nabla} \mathcal{S}'^-\Omega^n(*H) \longrightarrow 0$$

$$(2.9)\quad \mathcal{S}_{\hat{H}}^-(*H) \xrightarrow{\nabla} \mathcal{S}_{\hat{H}}^-\Omega^1(*H) \xrightarrow{\nabla} \cdots \xrightarrow{\nabla} \mathcal{S}_{\hat{H}}^-\Omega^n(*H) \longrightarrow 0$$

$$(2.10)\quad \mathcal{S}_0^- \xrightarrow{\nabla} \mathcal{S}_0^-\Omega^1 \xrightarrow{\nabla} \cdots \xrightarrow{\nabla} \mathcal{S}_0^-\Omega^n \longrightarrow 0 .$$

Suppose that the following: for any point $p \in H$, there exist a neighborhood U with holomorphic coordinates $x_1,\ldots,x_n$, $U \cap H = \{x_1 \cdots x_{n''} = 0\}$ and a free basis $e_U = \langle e_{1U},\ldots,e_{mU}\rangle$ for S(*H) over U such that the connection matrix relative to e_U and $x_1,\ldots,x_n$

$$\Omega_{e_U} = \textstyle\sum_{i=1}^{n''} x^{-p_i} x_i^{-1} A_i(x)dx_i + \sum_{i=n''+1}^{n} x^{-p_i} A_i(x)dx_i ,$$

satisfies one of the following conditions (H.1) and (H.2):

(H.1) $p_i = 0$ and $A_i(0)$ has no eigenvalue of integer for all $i=1,\ldots,n''$,

or

(H.2) $p_{ii} > 0$ and $A_i(0)$ is invertible, or $p_i = 0$ and $A_i(0)$ has no eigenvalue of integer, for all i=1,...,n",

where $p_i=(p_{i1},\dots,p_{in''},0,\dots,0)\in\mathbb{N}^n$ and $A_i(x)$ is an m-by-m matrix of holomorphic functions in U for i=1,...,n.

By using the results of Part II, we can prove the following theorem.

<u>Theorem 2.1.</u> (Asymptotic ∇-Poincaré's Lemma)

<u>If the condition (H.1) is satisfied for any point on H, then the sequences (2.6), (2.7), (2.8), (2.9) and (2.10) are exact. If one of (H.1) and (H.2) is satisfied for any point on H, then the sequences (2.6), (2.7) and (2.10) are exact.</u>

Proof of Theorem 2.1. At a point $p'\notin pr^{-1}(H)$, the stalks are

$$(\mathcal{S}^{-}\Omega^q(*H))_{p'} = (\mathcal{S}'^{-}\Omega^q(*H))_{p'} = (\mathcal{S}_0^{-}\Omega^q)_{p'} = (\mathcal{S}\Omega^q(*H))_{pr(p')} \cong ((\Omega^q)_{pr(p')})^m$$

and

$$(\mathcal{S}eg^{-}\Omega^q(*H))_{p'} = (\mathcal{S}_{\hat{H}}^{-}\Omega^q(*H))_{p'} = 0 .$$

and the results follows from the holomorphic ∇-Poincaré's Lemma (see the fact noticed after Theorem B in Section IV.1) and the results are trivial, respectively.

At $p\in H$, we take a neighborhood U with holomorphic coordinates $x_1,\dots,x_n$ and choose a free basis $e_U=\langle e_{1U},\dots,e_{mU}\rangle$ such that the connection matrix satisfies (H.1) or (H.2). What must be verified are:

(*) Let u be a holomorphic q-form in an open polysector S(c,r) at p such that $(d-\Omega_{e_U})u=0$ there and that, for some $s=(s_1,\dots,s_{n''},0,\dots,0)\in\mathbb{N}^n$, $x^s u$ is strongly asymptotically developable there. Then, there exists a holomorphic (q-1)-form w in any sufficiently small open subpolysector S(c',r') of S(c,r) at p such that $(d-\Omega_{e_U})w=u$ in S(c',r') and that $x^s w$ is strongly asymptotically developable in S(c',r'). If u is strongly asymptotically developable to 0, then w can be taken as a (q-1)-form strongly asymptotically developable to 0. If (H.1) holds at p and if $x^s u$ is strongly asymptotically developable to $(\mathcal{O}_{M\hat{|}H})_p$, then $x^s w$ can be taken as a (q-1)-form strongly asymptotically developable to $(\mathcal{O}_{M\hat{|}N})_p$.

(**) Let F be a consistent family $\{u(q_J)\colon J\subset[1,n''],\ q_J\in\mathbb{N}^J\}$ of q-forms in an open polysector $S(c,r)$ such that $(d-\Omega_{e_U})(PS_J(F))=0$ for any subset J of $[1,n'']$. Then, there exists a consistent family G of (q-1)-forms in any sufficiently small open subpolysector $S(c',r')$ of $S(c,r)$ such that $(d-\Omega_{e_U})(PS_J(G))=PS_J(F)$ for any subset J of $[1,n'']$.

(***) Let u be a formal meromorphic ∇-closed q-form i.e. $u\in(\mathcal{O}_{M\hat{|}H}\otimes_{\mathcal{O}}\Omega^q(*H))_p)^m$ and $(d-\Omega_{e_U})u=0$. Suppose that (H.1) holds at p. Then, there exists a formal meromorphic (q-1)-form $w\in(\mathcal{O}_{M\hat{|}H}\otimes_{\mathcal{O}}\Omega^{q-1}(*H))_p)^m$ such that $(d-\Omega_{e_U})w=u$.

PROOF OF (*). Let n' be a positive integer inferior or equal to n and let u be of the following form: $u=\sum_{\#I=q,I\subset[n',n]}f_I dx_I$. Rewrite u in the following form: $u=dx_{n'}\wedge v_1+v_2$, where

$$v_1=\sum_{\#I=q,n'\in I}f_I dx_{I-\{n'\}},\qquad v_2=\sum_{\#I=q,n'\notin I}f_I dx_I\ .$$

Suppose that $(d-\Omega_{e_U})u=0$. Then, for h, $1\le h<n'$, and for I containing n', we obtain

$$((e_h\partial/\partial x_h)-x^{-p_h}A_h)f_I=0\ .$$

where $e_h=x_h$ for $h=1,\dots,n''$ and $e_h=1$ for $h=n''+1,\dots,n$. Then, this implies that the system

$$((e_{n'}\partial/\partial x_{n'})-x^{-p_{n'}}A_{n'})F_I=f_I\ ,$$

$$((e_h\partial/\partial x_h)-x^{-p_h}A_h)F_I=0\ ,\quad 1\le h<n'\ ,$$

satisfies the integrability condition. Choose $s=(s_1,\dots,s_{n''},0,\dots,0)\in\mathbb{N}^n$ such that $x^s f_I$ is strongly asymptotically developable in $S(c,r)$ and $\lim_{x\to 0}x^s f_I=0$ for all subset I of $[n',n]$, $n'\in I$. Put $F_I=x^{-s}G_I$, then the system is written in the following form

$$(S)^I \quad \begin{cases} x^{p_{n'}}((e_{n'}\partial/\partial x_{n'})-s_{n'})G_I = A_{n'}G_I+x^s f_I \ , \\ x^{p_h}((e_h\partial/\partial x_h)-s_h)G_I = A_h G_I \ , \quad 1\leq h < n' \ . \end{cases}$$

By Theorem II.4.2 or Theorem II.4.6, for any sufficiently small open subpolysector $S(c',r')$ of $S(c,r)$, the system $(S)^I$ has a solution G_I holomorphic and strongly asymptotically developable in $S(c',r')$. Moreover, if $x^s f_I$ is strongly asymptotically developable to $(\mathcal{O}_{M|\hat{H}})_p$ and to 0, then G_I is also strongly asymptotically developable to $(\mathcal{O}_{M|\hat{H}})_p$ and to 0, respectively. Set

$$\omega = \sum_{\#I=q,n'\in I} x^{-s} G_I dx_{I-\{n'\}} \ .$$

Then, we see easily that

$$u - (d-\Omega_{e_U})\,\omega = \sum_{\#I=q,I\subset[n'+1,n]} g_I dx_I \ .$$

Therefore, by an induction on n', we can construct the desired $(q-1)$-form w. Q.E.D. of (*).

PROOF OF (**). By the same argument as in the proof of (*), we reduce (**) to solve the following formal systems

$$\begin{cases} x^{p_{n'}}((e_{n'}\partial/\partial x_{n'})-s_{n'})PS_J(G_I) = A_{n'}PS_J(G_I)+PS_J(x^s F_I) \ , \\ x^{p_h}((e_h\partial/\partial x_h)-s_h)PS_J(G_I) = A_h PS_J(G_I) \ , \quad 1\leq h < n' \ , \end{cases}$$

where F_I is a given consistent family in $S(c,r)$ for $I\subset[n',n]$, $I\ni n'$, and J is any subset of $[1,n'']$. By the same argument in the proof of Theorem II.4.2 or Theorem II.4.6, we can assert that, for any sufficiently small open subpolysector $S(c',r')$ of $S(c,r)$, there exists a consistent family G_I in $S(c',r')$ such that for any subset J of $[1,n'']$, $PS_J(G_I)$ satisfies the above system. Q.E.D. OF (**).

PROOF OF (***). By the same argument as in the proof of (*), the assertion (***) is reduced to solve the following formal system

$$\begin{cases} x^{p_{n'}}((e_{n'}\partial/\partial x_{n'})s_{n'})G_I = A_{n'}G_I + x^s F_I \\ x^{p_h}((e_h\partial/\partial x_h)-s_h)G_I = A_h G_I , \quad 1 \leq h < n' , \end{cases}$$

where F_I is a given formal power-series vector in $((\mathcal{O}_{M\hat{|}H})_p)^m$ for $I \subset [n'.n]$, $I \ni n'$, and J is any subset [1,n"]. By the argument in the proof of Theorem II.4.2 (cf. Corrolary II.3.3), we can assert that there exists a formal power-series solution G in $((\mathcal{O}_{M\hat{|}H})_p)^m$ to the above system. Q.E.D. OF (***). Q.E.D. OF Theorem 2.1.

REMARK 2.1. By Remark II.3.1, we can prove Theorem 2.1 under the assumption that Ω is decomposable in the form $\oplus_{j=1}^{s}\Omega_j$ and each Ω_j satisfies (H.1) or (H.2).

SECTION IV.3. ∇-DE RHAM COHOMOLOGY THEOREM.

We keep notation used in the previous section. At first, we deduce some results from the asymptotic ∇-Poincaré's Lemma (Theorem 2.1).

PROPOSITION 3.1. Under the same assumption in Theorem 2.1,

1) the kernel sheaf $\mathcal{K}er\nabla^-$ of (2.1) is equal to the kernel sheaf $\mathcal{K}er\nabla_0^-$ of (2.5).

2) for i=0,1,2,..., we have the isomorphisms

$$(3.1)\quad \mathbb{H}^i(M^-,(\mathcal{S}_0^-\Omega^\cdot,\nabla_0^-)) = H^i(M^-,\mathcal{K}er\nabla_0^-) = \mathbb{H}^i(M^-,(\mathcal{S}^-\Omega^\cdot(*H),\nabla^-)) .$$

PROOF OF 1). By the assumption, for any point $p\in H$, the formal equation

$$(d-\Omega_{e_U})u = 0$$

has only one solution which is equal to zero. So, a strongly asymptotically developable solution to the system $(d-\Omega_{e_U})u=0$ is strongly asymptotically developable to 0. Q.E.D.

PROOF OF 2). In general, let $(K^\cdot,\delta)$ be a complex of sheaves over M^-

$$(K^\cdot,\delta) :\quad K_1 \xrightarrow{\delta} K_2 \xrightarrow{\delta} \cdots \xrightarrow{\delta} K_n \longrightarrow 0.$$

If "Poincaré's lemma holds for this complex $(K^\cdot,\delta)$, then we have the isomorphisms

$$\mathbb{H}^i(M^-,(K^\cdot,\delta)) = H^i(M^-, \mathcal{K}er\,\delta) ,$$

for i=0,1,2,..., where $\mathcal{K}er\,\delta$ is the kernel sheaf of $\delta: K_1\longrightarrow K_2$. Therefore, we obtain the above isomorphism (3.1) from Theorem 2.1 and the assertion 1). Q.E.D.

By the same argument, we have the following.

PROPOSITION 3.2. Suppose that for a point p on H, a connection matrix of ∇ at p satisfies the condition (H.1) or (H.2). Then, we have the isomorphisms

$$(3.2)\quad \mathbb{H}^i(p^-,(\mathcal{S}_0^-\Omega^\cdot|_{p^-},\nabla_0^-|_{p^-})) = H^i(p^-,\mathcal{K}er\nabla_0^-|_{p^-}) = \mathbb{H}^i(p^-,(\mathcal{S}^-\Omega^\cdot(*H)|_{p^-},\nabla^-|_{p^-})) ,$$

for $i=0,1,2,\ldots,$ *where* $p^-=pr^{-1}(p)$.

LEMMA 3.1. *Under the same assumption as in Theorem 2.1, we have a natural homomorphism*

$$\alpha^i : H^i(\Gamma(M^-,\mathcal{S}^-\Omega^\cdot(*H)),\nabla^-)\longrightarrow H^i(M^-,\mathcal{K}er\nabla_0^-) \ ,$$

for $i=0,1,2,\ldots$

PROOF. In the case of $i=0$, this assertion is trivial.

In the case of $i=1$, let $\omega\in\Gamma(M^-,\mathcal{S}^-\Omega^1(*H))$, $\nabla\omega=0$ and let $\{U_k^-\}_{k\in K}$ be an open covering of M^-. If U_k^-'s are sufficiently small, then, by Theorem 2.1, we obtain a local section $u_k\in\Gamma(U_k^-,\mathcal{S}^-(*H))$ satisfying $\nabla u_k=\omega$ in U_k^- for all $k\in K$. Put $u_{kk'}=u_{k'}-u_k$ for $k, k'\in K$. Then, we see easily $u_{kk'}=0$ in $U_k^-\cap U_{k'}^-$ and $u_{kk'}\in\Gamma(U_k^-\cap U_{k'}^-,\mathcal{S}^-(*H))$. Moreover, by Proposition 3.1, $u_{kk'}\in\Gamma(U_k^-\cap U_{k'}^-,\mathcal{S}_0^-)$, i.e., $u_{kk'}$ is strongly asymptotically developable to 0. Hence, the family $\{u_{kk'}\}$ becomes a 1-cocycle of the covering with coefficients in the sheaf $\mathcal{K}er\nabla_0^-$. Denote by $[\{u_{kk'}\}]$ the class of cohomology in $H^1(M^-,\mathcal{K}er\nabla_0^-)$. Then, this class is well-defined. We shall check it. If u_k' is another solution to the equation $\nabla u_k'=\omega$ in U_k^-, then $\{u_{kk'}'=u_{k'}'-u_k'\}$ represents evidently the same class as $\{u_{kk'}\}$. If $\omega'=\omega+\nabla z$ with a global section $z\in\Gamma(M^-,\mathcal{S}^-(*H))$ and if $\nabla v_k=\omega'$ in U_k^-, then $\{v_{kk'}=v_{k'}-v_k\}$ represents the same class as $\{u_{kk'}\}$: because $\nabla(u_k-v_k+z)=0$, we see, by Proposition 3.1 again, that $u_k-v_k+z\in\Gamma(U_k^-,\mathcal{S}_0^-)$ and obviously

$$u_{kk'}-(u_{k'}-v_{k'}+z)=v_{kk'}-(u_k-v_k+z)$$

in $U_k^-\cap U_{k'}^-$. Therefore, we can define $\alpha^1(\omega)=[\{u_{kk'}\}]$.

In the case of $i=2$, let $\theta\in\Gamma(M^-,\mathcal{S}^-\Omega^2(*H))$ and let $\{U_k^-\}_{k\in K}$ be the open covering of M^- as above. If $\nabla\theta=0$, then by Theorem 2.1, there exists a $\omega_k\in\Gamma(U_k^-,\mathcal{S}^-\Omega^1(*H))$ satisfying $\nabla\omega_k=\theta$ for any $k\in K$. Moreover, $\nabla(\omega_{k'}-\omega_k)=0$, there exists $u_{kk'}\in\Gamma(u_k^-\cap U_{k'}^-,\mathcal{S}^-\Omega^1(*H))$ such that $\nabla u_{kk'}=\omega_{k'}-\omega_k$. Put $u_{kk'k''}=u_{kk'}+u_{k'k''}+u_{k''k}$. Then, we see easily $\nabla u_{kk'k''}=0$ and so $u_{kk'k''}$

$\in \Gamma(U_k^- \cap U_{k'}^- \cap U_{k''}^-, \mathcal{S}_0^-)$. Thus, we obtain a 2-cocycle $\{u_{kk'k''}\}$ of the covering with coefficients in the sheaf $\mathcal{K}er\,\nabla_0^-$, and we can easily verify that the cohomology class $[\{u_{kk'k''}\}]$ is well defined. Therefore, we can define $\alpha^2(\theta)=[\{u_{kk'k''}\}]$.

In the case of $i \geq 3$, by the same argument as above, we can define α^i. Q.E.D.

By the same argument, we deduce the following.

LEMMA 3.2. For a point p on H, we suppose that (H.1) or (H.2) are satisfied. Then, we have a natural homomorphism

$$\alpha_p^i : H^i(\Gamma(p^-, \mathcal{S}^-\Omega^\cdot(*H)|_{p^-}), \nabla^-|_{p^-}) \longrightarrow H^i(p^-, \mathcal{K}er\,\nabla_0^-|_{p^-}),$$

for $i=0,1,2,\ldots$, where $p^-=pr^{-1}(p)$.

In order to construct the inverse homomorphisms, we need the vanishing theroem.

THEOREM 3.1. Let $\mathcal{S}$ be a locally free sheaf over M of $\mathcal{O}$-modules or $\mathcal{O}(*H)$-modules of rank m.

i) For any point $p \in H$, the natural homomorphism

$$H^1(p^-, \mathcal{S}_0^-\Omega^q|_{p^-}) \longrightarrow H^1(p^-, \mathcal{S}^-\Omega^q|_{p^-})$$

is a zero mapping, for $q=0,1,2,\ldots$, where $p^-=pr^{-1}(p)$ and $\mathcal{S}^-\Omega^q=pr^*(\mathcal{S}\otimes_{\mathcal{O}}\Omega^q)$.

2) If $H^1(M,(\mathcal{S}\otimes_{\mathcal{O}}\Omega^q))=0$, then the natural homomorphism

$$H^1(M^-, \mathcal{S}_0^-\Omega^q) \longrightarrow H^1(M^-, \mathcal{S}^-\Omega^q)$$

is a zero mapping for $q=0,1,2,\ldots$

The proof of this theorem is the same as those of Theorems I.3.2 and I.3.3 and so omitted.

REMARK 3.1. In general, $H^q(p^-, \mathcal{S}_0^-\Omega^q|_{p^-})$ and $H^q(M^-, \mathcal{S}_0^-\Omega^q)$ have complicate structures for $q \geq 2$. In the simple case where H is a non-singular hyperplane, we see easily that $H^q(P^-, \mathcal{S}_0^{-}\Omega^q|_{P^-})=0$ for $q \geq 2$, because of existence of a "good" open covering.

<u>LEMMA 3.3</u>. <u>If</u> $H^1(M, \mathcal{S})=0$, <u>then there exists a natural homomorphism</u>

$$\beta^1 : H^1(M^-, \mathcal{K}er\nabla_0^-) \longrightarrow H^1(\Gamma(M^-, \mathcal{S}^-\Omega^{\cdot}(*H)), \nabla^-) .$$

PROOF. Let $\{U_k^-\}_{k\in K}$ be an open covering of M^- and let $\{u_{kk'}\}$ be a 1-cocycle of the covering with coefficients in the sheaf $\mathcal{K}er\,\nabla_0^-$. Suppose that the covering is sufficiently fine. Then, by applying Theorem 3.1.2) to this 1-cocycle, we obtain a 0-cochain $\{u_k\}_{k\in K}$ of the covering with coefficients in $\mathcal{S}^-$ such that

$$u_{kk'} = u_{k'} - u_k \quad \text{in} \quad U_k^- \cap U_{k'}^- .$$

Because of $\nabla u_{kk'}=0$, we see

$$\nabla u_{k'} = \nabla u_k \quad \text{in} \quad U_k^- \cap U_{k'}^- .$$

Therefore, we can define $\omega \in \Gamma(M^-, \mathcal{S}^-\Omega^1(*H))$ by

$$\omega = \nabla u_k \quad \text{in} \quad U_k^-, \ k\in K .$$

Denote by $[\omega]$ the class of cohomology of ω in $H^1(\Gamma(M^-, \mathcal{S}^-\Omega^{\cdot}(*H)), \nabla^-)$. We shall verify that this class is uniquely determined by the class $[\{u_{kk'}\}]$ of cohomology in $H^1(M^-, \mathcal{K}er\nabla_0^-)$. Let $\{v_{kk'}\}$ be another 1-cocycle of the covering with coefficients in $\mathcal{K}er\,\nabla_0^-$. By the same argument as above, there exists a 0-cochain $\{v_k\}_{k\in K}$ such that $v_{kk'}=v_{k'}-v_k$ in $U_k^-\cap U_{k'}^-$. Suppose that the 1-cocycle $\{v_{kk'}\}$ respresents the same class of cohomology as $\{u_{kk'}\}$. Then, there exists a 0-cochain $\{w_k\}_{k\in K}$ of the covering with coefficients in $\mathcal{K}er\,\nabla_0^-$ such that

$$u_{kk'} = v_{kk'} - w_k + w_{k'} \text{ in } U_k^- \cap U_{k'}^-,$$

from which we see that

$$u_{k'} - v_{k'} - w_{k'} = u_k - v_k - w_k \quad \text{in} \quad U_k^- \cap U_{k'}^- .$$

Define $z \in \Gamma(M^-, \mathcal{S}^-)$ by

$$z = u_k - v_k - w_k \quad \text{in} \quad U_k^-, \ k\in K .$$

Then, we obtain the equality $\nabla u_k = \nabla v_k + \nabla z$ in U_k^- for all $k \in K$. This implies $\omega = \omega' + \nabla z$, where $\omega' = \nabla v_k$ in U_k^- for $k \in K$, i.e. $[\omega] = [\omega']$. Hence, we can define β^1 by $\beta^1(]\{u_{kk'}\}]) = [\omega]$. Q.E.D.

By the same argument, we can prove

LEMMA 3.4. *For any point* $p \in H$, *there exists a natural homomorphism*

$$\beta_p^1 : H^1(p^-, \mathcal{K}er\,\nabla_0^-|_{p^-}) \longrightarrow H^1(\Gamma(p^-, \mathcal{S}^-\Omega^\cdot(*H))|_{p^-}, \nabla^-|_{p^-}) ,$$

where $p^- = pr^{-1}(p)$.

Now, we give an analogue of de Rham cohomology theorem.

THEOREM 3.2. *If* $H^1(M, \mathcal{S}) = 0$ *and if* (H.1) *of* (H.2) *are satisfied for any point on* H, *then we have a natural isomorphism*

$$b^1 : H^1(M^-, \mathcal{K}er\,\nabla_0^-) \longrightarrow H^1(\Gamma(M, \mathcal{S}\Omega^\cdot(*H)), \nabla) .$$

PROOF. By the definitions of α^1 and β^1 in Lemma 3.1 and Lemma 3.3, respectively, we see easily that $\alpha^1 \cdot \beta^1$ and $\beta^1 \cdot \alpha^1$ are identity mapping, respectively. Hence, α^1 and β^1 are isomorphisms. On the othre hand, from the definition, we obtain

$$H^1(\Gamma(M, \mathcal{S}\Omega^\cdot(*H)), \nabla) = H^1(\Gamma(M^-, \mathcal{S}^-\Omega^\cdot(*H)), \nabla^-) . \quad \text{Q.E.D.}$$

By Lemmas 3,2 and 3.4, we also obtain

THEOREM 3.3. *For any point* p *on* H, *if the assumption as in Theorem 2.1 is satisfied at* p, *then we have a natural isomorphism*

$$b_p^1 : H^1(p^-, \mathcal{K}er\,\nabla_0^-|_{p^-}) \longrightarrow H^1((\mathcal{S}\Omega^\cdot(*H))_p, \nabla) ,$$

where $p^- = pr^{-1}(p)$ *and* $(\mathcal{S}\Omega^\cdot(*H))_p$ *is the stalk of the sheaf* $\mathcal{S}\Omega^\cdot(*H)$ at p.

References

[1] Baldassarri, F.: Differential module and singular points of p-adic differential equations, Advances in Math., 44, 155-179 (1982).

[2] Balser, W., Jurkat, B. and Lutz, D.A.: Birkhoff invariants and Stokes multipliers for meromorphic linear differential equations, J. Math. Anal. Appl., 71, 48-94 (1979).

[3] Balser, W., Jurkat, B. and Lutz, D.A.: A general theory of invariants for meromorphic differential equations; Part I, formal invariants; Part II, proper invariants, Funk. Ekva., 22, 197-221, 257-283 (1979); Part III, Houston J. Math., 6, 149-189 (1980).

[4] Birkhoff, G.D.: Singular points of ordinary linear differential equations, Trans. Amer. Math. Soc., 10, 436-470 (1909).

[5] Birkhoff, G.D.: The generalized Riemann problem for linear differential equations and the allied problems for linear difference equations, Proc. Amer. Acad. Arts and Sci., 49, 521-568 (1913).

[6] Brieskorn, E.: Die monodromie der isolierten singularitäten von hyperflächen, Manuscripta Math., 2, 103-161 (1970).

[7] Charrière, H.: Triangulation formelle de certains systèmes de Pfaff complètement intégrables et application à l'étude C^{∞} des systèmes lineáires. Publ. I.R.M.A., Strasbourg (1980).

[8] Charrière, H. and Gérard, R.: Formal reduction of integrable linear connections having a certain kind of irregular singularities, Analysis, 1, 85-115 (1981).

[9] Coddington, E. and Levinson, N.: Theory of ordinary differential equations, McGraw-Hill, New York (1955).

[10] Cope, F.T.: Formal solutions of irregular linear differential equations, I, II, Amer. J. Math., 56, 411-437 (1934); 58, 130-140 (1936).

[11] Deligne, P.: Equations differentielles a points singuliers reguliers, Lecture Notes in Math., 163, Springer-Verlag (1970) and Correction to Lecture Note 163, Warwick University, April 1971.

[12] Fabry, E.: Sur les intégrales des equations différentielles linéaires à coefficients rationnels, Thèse Paris, 1885.

[13] Gérard, R.: Théorie de Fuchs sur une variété analytique complexe, J. Math. Pures et Appl., 47, 321-404 (1968).

[14] Gérard, R.: Le probleme de Riemann-Hilbert sur une varieté analytique complexe, Ann. Inst. Fourier, 19, 1-12 (1969).

[15] Gérard, R. and Levelt, A.: Sur les connections à singularités réguliers dans le cas de plusieurs variables, Funk. Ekva., 19, 149-173 (1976).

[16] Gérard, R. and Levelt, A.H.M.: Invariants mesurant l'irrégularité en un point singulier des systèmes d'équations différentielles linéaires. Ann. Inst. Fourier, 23, 157-195 (1973).

[17] Gérard, R. and Sibuya, Y.: Etude de certains systèmes de Pfaff avec singularities, Lecture Notes in Math., 712, 131-288, Springer-Verlag (1979).

[18] Giraud, J.: Cohomologie non abélienne, Grund. Math. Wiss., 179, Springer-Verlag (1971).

[19] Grauert, H. and Remmert, R.: Theory of Stein spaces, Grund. Math. Wiss., 236, Springer-Verlag (1979).

[20] Griffiths, P.: Periods of integrals on algebraic manifolds, Bull. Amer. Math. Soc., 76, 228-296 (1970).

[21] Griffiths, P. and Harris, J.: Principles of algebraic geometry, Pure and Applied Math., Wiley-Interscience Pub., J. Wiley and Sons, New York (1978).

[22] Grothendieck, A.: On the de Rham cohomology of algebraic varieties, Publ. Math. I.H.E.S., 29, 95-103 (1966).

[23] Harris, W.A. Jr.: Analytic theory of linear differential systems, Lecture Notes in Math., 243, 229-237, Springer-Verlag (1971).

[24] Hironaka, H.: Resolution of singularities of an algebraic variety over a field of characteristic zero I, II, Ann. Math., 79, 109-326 (1964).

[25] Hitotsumatsu, S.:Theory of analytic functions of several variables, Baifukan, Tokyo (1960) (in Japanese).

[26] Hsieh, P.-F.: Regular perturbation for a turning point problem, Funk. Ekva., 12, 155-179 (1969).

[27] Hukuhara, M.: Sur les points singuliers des équations différentielles linéaires, II, Jour. Fac. Sci. Hokkaido Univ., 5, 123-166 (1937).

[28] Hukuhara, M.: Sur les points singuliers des équations différentielles linéaires, III, Mem. Fac. Sci. Kyushu Univ., 2, 125-137 (1942).

[29] Ince, E.L.: Ordinary differential equations, Dover, New York (1944).

[30] Iwano, M.: Bounded solutions and stable domains of nonlinear ordinary differential equations, Lect. Notes in Math., 183, 59-127, Springer-Verlag (1971).

[31] Jurkat, W.B.: Meromorphe differentialgleichungen, Lect. Notes in Math., 637, Springer-Verlag (1967).

[32] Jurkat, W.B. and Lutz, D.A.: On the order of solutions of analytic differential equations, Proc. London Math. Soc., 22, 465-482 (1971).

[33] Jurkat, W.B., Lutz, D.A. and Peyerimhoff, A.: Birkhoff invariants and effective calculations for meromorphic linear differential equations, I, J. Math. Anal. Appl., 53, 438-470 (1976); II, Houston J. Math., 2, 207-238 (1976).

[34] Katz, N.: Nilpotent connections and the monodromy theorem: applications of a result of Turrittin, Publ. Math. I.H.E.S., 39, 175-232 (1970).

[35] Katz, N.: An overview of Deligne's work on Hilbert's twenty-first problem, Proc. Symp. in Pure Math. Vol.28, 537-557 (1976).

[36] Kimura, T.: Hypergeometric functions of two variables, Lect. Note, University of Tokyo (1973).

[37] Kita, M.: The Riemann-Hilbert problem and its application to analytic functions of several variables, I, II, Tokyo J. Math., 2, 1-27; 293-300 (1979).

[38] Kitagawa, K.: Charactérization de singularité régulier, preprint (1981), to appear, Jour. of Univ. of Kyoto.

[39] Kohno, M. and Okubo, K.: Asymptotic expansions, Kyoikushuppan, Tokyo (1976) (in Japanese).

[40] Levelt, A.: Jordan decomposition of a class of singular differential operators, Arkiv. for Math., 13, 1-27 (1975).

[41] Lin, C.-H.: Phragman-Lindelöf theorem in a cohomological form, to appear in Proc. Amer. Math. Soc.

[42] Lin, C.-H.: The sufficiency of Matkowski-condition in the problem of resonance, Thesis, University of Monnesota (1982); to appear in Trans. Amer. Math. Soc.

[43] Majima, H.: Remarques sur la théorie de developpement asymptotique de plusieurs variables, I, Proc. Jap. Acad., 54, 67-72 (1978).

[44] Majima, H.: On reduced systems of the Pfaffian systems of confluent hypergeometric functions of two variables (private note).

[45] Majima, H.: Sur les systemes de Pfaff complètement integrable dont les coefficients ont des singularitès au plus sur les axes de $\mathbb{C}^n$, preprint (1978).

[46] Majima, H.: On Pfaffian systems with singularities, Proc. Sem. (Kokyuroku) at R.I.M.S., Univ. of Kyoto, No. 351, 1-13 (1979) (in Japanese).

[47] Majima, H.: Etudes sur les systèmes d'equations différentielles aux derivées partielles du premier ordre à points singuliers réguliers dan le cadre de developpement asymptitque; ....singuliers irreguliers...., preprint (1981).

[48] Majima, H.: On the representation of solutions of completely integrable Pfaffian systems with irregular singular points, Proc. Sem. at R.I.M.S. (Kokyuroku), Kyoto Univ., No.438 (1981) (in Japanese).

[49] Majima, H.: Analogues of Cartan's decomposition theorems in asymptotic analysis, preprint (1982), to appear in Funk. Ekva.

[50] Majima, H.: Vanishing theorems in asymptotic analysis, Proc. Japan Acad., 59, Ser. A, 146-149 (1983).

[51] Majima, H.: ∇-Poincaré's lemma and an isomorphism theorem of de Rham type in asymptotic analysis, preprint (1983).

[52] Majima, H.: ∇-Poincaré's lemma and ∇-de Rham cohomology for an integrable connection with irregular singular points, Proc. Japan Acad., 59, Ser. A, 150-153 (1983).

[53] Majima, H.: Riemann-Hilbert-Birkhoff problem for integrable connections with irregular singular points, Proc. Japan Acad., 59, Ser. A, (1983).

[54] Majima, H.: Integrable connections with irregular singular points and the Riemann-Hilbert-Birkhoff problem, preprint (1983).

[55] Malgrange, B.: Sur les points singuliers des équations différentielles, l'Enseignment Math., 20, 147-176 (1974).

[56] Malgrange, B.: Remarques sur les équations différentielles à points singuliers irréguliers, Lect. Notes in Math., 712, 77-86, Springer-Verlag (1979).

[57] Malgrange, B.: Sur le reduction formelles des équations différentielles à singuliers irréguliers, preprint (1979).

[58] Malmquist, J.: Sur l'études analytique des solutions d'un système des équations différentielles dans le voisinage d'un point singulier d'indétermination, I; II; III, Acta Math., 73, 87-129 (1940); 74, 1-64, 109-128 (1941).

[59] Manin, J.: Moduli fuchsiani, Ann. Sc. Norm. Sup. Pisa, 19, 113-126 (1965).

[60] Martinet, J. and Ramis, J.-P.: Problèmes de modules pour des équations différentielles non-linéaires due premier ordre, Publ. I.H.E.S., 55, 63-164 (1982).

[61] Moser, J.: The order of a singularity in Fuchs' theory, Math. Zei., 72, 379-398 (1960).

[62] Nakano, Y.: Theory of functions of several variables, Math. Sci. Lib., 4, Asakura-shoten (1982) (in Japanese).

[63] Nilsson, N.: Some growth and ramification properties of certain integrals on algebraic manifolds, Arkiv. for Math., 5, 527-540 (1963-65).

[64] Poincaré, H.: Sur les intégrales des équations linéaires, Acta Math., 8, 295-344 (1886).

[65] Ramis, J.-P.: Devissage Gevrey, Asterisque, 5960, 173-204 (1978).

[66] Ramis, J.-P.:Théorèmes d'indices Gevrey pour les équations différentielles ordinaires, Publ. I.R.M.A., Strasbourg (1981).

[67] Robba, P.: Lemmes de Hensel pour les operateurs differentials. Application à la reduction formelle des équations différentielles, Ens. Math., 26, 279-311 (1980).

[68] Röhrl, H.: Das Riemann-Hilbertsche Problem der Theorie der linearen differentialgleichungen, Math. Ann., 133, 1-25 (1957).

[69] Sibuya, Y.: Simplification of a system of linear ordinary differential equations about a singular point, Funk. Ekva., 4, 29-56 (1962).

[70] Sibuya, Y.: Perturbation of linear ordinary differential equations at irregular singular points, Funk. Ekva., 11, 235-246 (1968).

[71] Sibuya, Y.: Perturbation at an irregular singular point, Lect. Notes in Math., 243, 148-168, Springer-Verlag (1971).

[72] Sibuya, Y.: Global theory of second order linear ordinary differential equations with a polynomial coefficient, Math. Studies 18, North-Holland (1975).

[73] Sibuya, Y.: Linear ordinary differential equations in the complex domain-connection problems-, Kinokuniya-shoten (1976) (in Japanese).

[74] Sibuya, Y.: Stokes phenomena, Bull. Amer. Math. Soc., 83, 1075-1077 (1977).

[75] Sibuya, Y.: Convergence of power series solutions of a linear Pfaffian system at an irregular singularity. Keio Engineering Reports, Vol. 31, 79-86 (1978); A linear Pfaffian system at an irregular singularity, Tohoku Math. Jour., 32, 209-215 (1980).

[76] Sibuya, Y.: A theorem concerning uniform simplification at a transition point and the problem of resonance, SLAM J. Math. Anal., 12, 653-668 (1981).

[77] Suzuki, O.: The problem of Riemann and Hilbert and the relations of Fuchs in several complex variables, Lecture Notes in Math., 712,.325-364, Springer-Verlag (1979).

[78] Takano, K,: Asymptotic solutions of linear Pfaffian systems with irregular singular points, Jour. Fac. Sci. Sec. IA, 24, 381-404 (1977).

[79] Takano, K. and Yoshida, M.: On a linear system of Pfaffian equations with regular singular points, Funk. Ekva., 19, 147-176 (1976).

[80] Trijitzinsky, W.J.: Analytic theory of linear differential equations, Acta Math., 62, 167-226 (1933).

[81] Turrittin, H.L.: Asymptotic expansions of solutions of systems of ordinary linear differential equations containing a parameter, Ann. Math., 29 (Contribution to the theory of nonlinear oscillations, ed. by S. Lefschetz), 81-116, Princeton (1952).

[82] Turrittin, H.L.: Convergent solutions of ordinary homogeneous differential equations in the neighborhood of a singular point, Acta Math., 93, 27-66 (1955).

[83] Wasow, W.: Asymptotic expansions of ordinary differential eauations, Interscience (1965); R.E. Krieger Pub. Comp. (1976).

[84] Van den Essen, A. and Levelt, A.: Irregular singularities in several variables, Memoirs Amer. Math. Soc. Vol. 40, No. 270 (1982).

[85] Sibuya, Y. and Majima, H.: Cohomological characterization of regular singularity in several variables , preprint (1984)

[86] Majima, H.: Vanishing theorems in asymptotic anlysis II, Proc. Japan Acad., 60 Ser. A, 171-173 (1984).

Subject Index